What Your Colleagues Are Saying . . .

Whose Math Is It? is a must-read for any math teacher who seeks to create a classroom learning environment in which their students are actively engaged in challenging mathematics, thoughtfully and respectfully discussing their mathematical thinking, and personally aware of their own learning and agency. Whether you are new to or experienced with active learning, this book offers a wealth of concrete strategies that will expand and enrich your instructional repertoire. A book like this with such experience-based insights is a treasure and does not come along very often. I highly recommend it!

—**Chris Rasmussen**, PhD in mathematics education

Whose Math Is It? is about teaching students how to take ownership of "their math." A comprehensive book that includes examples, tools, strategies, and resources, it is great for any educator (teacher, co-teacher, instructional coach, or administrator) who wants all their students to see themselves as mathematicians who critically think about mathematics and talk about it in meaningful ways. Educators who want their math students to feel confident and believe in themselves will want to read this book!

—**Staci Benak**, EdD, math resource teacher, San Diego Unified School District

Student efficacy in the math class is attainable and should be a goal for every math teacher. *Whose Math Is It?* provides effective strategies to move the focus from teachers doing the heavy lifting to students becoming empowered in their learning. Joseph Michael Assof's book guides teachers in the creation of classroom systems that support student agency in learning.

—**Kim West**, Corwin faculty member, Kramer IB World School PYP coordinator, and math instructional coach, Dallas ISD

Whose Math Is It?

Whose Math Is It?

Building Student Ownership in Mathematics

Joseph Michael Assof

Foreword by Douglas Fisher

FOR INFORMATION:

Corwin

A SAGE Company

2455 Teller Road

Thousand Oaks, California 91320

(800) 233-9936

www.corwin.com

SAGE Publications Ltd.

1 Oliver's Yard

55 City Road

London EC1Y 1SP

United Kingdom

SAGE Publications India Pvt. Ltd.

Unit No 323-333, Third Floor, F-Block

International Trade Tower Nehru Place

New Delhi 110 019

India

SAGE Publications Asia-Pacific Pte. Ltd.

18 Cross Street #10-10/11/12

China Square Central

Singapore 048423

Vice President and
   Editorial Director: Monica Eckman

Director and Publisher: Lisa Luedeke

Content Development Editor: Sarah Ross

Product Associate: Zachary Vann

Production Editor: Tori Mirsadjadi

Copy Editor: Michelle Ponce

Typesetter: C&M Digitals (P) Ltd.

Proofreader: Eleni Maria Georgiou

Indexer: Integra

Cover Designer: Janet Kiesel

Marketing Manager: Megan Naidl

Printed in the United States of America

Library of Congress Cataloging-in-Publication Data

Names: Assof, Joseph, author. | Fisher, Douglas, 1965-writer of foreword.

Title: Whose math is it? : building student ownership in mathematics / Joseph Michael Assof ; foreword by Douglas Fisher.

Description: Thousand Oaks, California : Corwin, [2025] | Includes bibliographical references and index.

Identifiers: LCCN 2024008460 | ISBN 9781071949511 (paperback) | ISBN 9781071953860 (epub) | ISBN 9781071953877 (epub) | ISBN 9781071953884 (pdf)

Subjects: LCSH: Mathematics—Study and teaching. | Academic achievement. | Self-efficacy.

Classification: LCC QA11.2 .A874 2025 | DDC 510.71—dc23/eng/20240412

LC record available at https://lccn.loc.gov/2024008460

This book is printed on acid-free paper.

SUSTAINABLE FORESTRY INITIATIVE   Certified Sourcing   www.sfiprogram.org   SFI-01028

24 25 26 27 28 10 9 8 7 6 5 4 3 2 1

Contents

 Visit the companion website at
**https://companion.corwin.com/courses/
whosemathisit**
for downloadable resources.

Note From the Publisher: The author of this book has provided content that is available to you through a QR (quick response) code. To read a QR code, you must have a smartphone or tablet with a camera. We recommend that you download a QR code reader app that is made specifically for your phone or tablet brand.

Foreword

By Douglas Fisher

I have had the opportunity to observe Joseph Assof teaching on hundreds of occasions. He is an expert, skilled at guiding students' mathematical understanding, building their confidence and competence. Students leave his classes with more than procedural knowledge. They gain conceptual understanding and can apply what they have learned to novel situations. In other words, they reach the level of transfer, generalization, application, and authentic use of their knowledge.

Part of Joseph's belief about learning, which he has shared with teachers across the country, is that students have to develop ownership of, and responsibility for, their mathematics learning. This book is aptly titled *Whose Math Is It?* because Joseph knows that students must develop efficacy in their mathematical learning if they are to take responsibility for their learning. Students must have goals and align their efforts along with their goals. And they must experience the fruits of their labors, knowing that they are learning more and better as a result of their efforts.

This is no small feat. Far too many students approach mathematics with the belief that they will fail, that math is for other people, and that they are not capable of learning the content. We all recognize this fallacy, but Joseph shows us how to teach students that this is not reality; that they own their mathematical learning.

To accomplish this, Joseph embraces teacher clarity. In fact, he is one of the original authors of the *Teacher Clarity Playbook*, a book that outlines a process for analyzing standards and designing learning experiences. But teacher clarity is more than learning intentions and success criteria. It's also the meaningful experiences students have with the content.

Clarity requires that educators match instructional approaches with the appropriate phase of learning. Importantly, these phases can occur within a single lesson or across multiple lessons. Surface learning is not superficial, it's foundational or introductory. And there are tools teachers use to build students' surface learning. But we can't leave students there. When teachers change instructional approaches and tasks, they can move students to deep learning, during which time students make connections, see relationships, and develop schemata. Our goal is not adult-dependent learners but rather students who self-regulate and continue learning. At the transfer level, students can apply their learning in new situations. As I have noted before, it's the right approach, at the right time, for the right kind of learning.

Transfer of learning, the goal of our collective efforts, is not easy. In fact, the American Psychological Association (2015) notes that "student transfer or generalization of their knowledge and skills is not spontaneous or automatic" (p. 10). And that's the magic of this book. By teaching students that the math is theirs, that they own it and use it, students begin to transfer their learning. Joseph shows that there are processes and procedures that teachers can use to guide students' thinking without telling them what to think. Filled with examples across the grade levels, this book supports educators in developing a mindset with students that they are the owners of mathematics, that they can use math to solve interesting problems, and that they are responsible for their own learning.

About the Author

Joseph Michael Assof is a high school and community college mathematics teacher and the math department chair at Health Sciences High and Middle College in San Diego, CA. He is also an educational consultant and presents internationally on a wide array of topics including teacher clarity, mathematics teaching and learning, visible learning, and more. Joseph coauthored *Teaching Mathematics in the Visible Learning Classroom, High School*, *Teaching Mathematics in the Visible Learning Classroom, Grades 6–8*, and *The Teacher Clarity Playbook*, and his classroom is featured in a number of *Visible Learning for Mathematics, Grades K–12* videos. Joseph holds bachelor's and master's degrees in mathematics and a doctorate in Educational Leadership with an emphasis in Mathematics Teacher Leadership. Mathematics and mathematics education are Joseph's second passion—his first being his two beautiful boys, Joseph Fred and Jamie Beau.

Introduction

Keeping the End in Mind

Imagine a mathematics classroom where students are not only actively engaged in critical thinking, problem solving, and constructive argumentation, but where they are also aware of their own learning, seek feedback on their work, provide feedback to their peers, and monitor their own progress. One can imagine further that there exists a culture in this classroom of high expectations sustainable only by its equally prominent culture of support.

This is a classroom fueled by efficacy—where students are choosers and users of learning strategies that have proven effective for them in the past and thus give them confidence to use them again. This is a classroom where the teacher may truly embody the role of facilitating learning, with confidence that their expertise is not going underutilized. Now, compare this abstract ideal to the concrete reality. This comparison might tempt some down the student-by-student road; checking off individual talented students who could rise to the occasion of such an idealized classroom and crossing off others who likely would not. This approach, however, begs the question: Do we develop or select talent? And while many of us in education might instinctually and fervently (and commendably) react to such a question, without efficacy of our own, the prospect of developing such a high degree of talent might seem unattainable.

Thus is the purpose of this text. This book seeks to act as the representational intermediary between the abstract ideal classroom described above and the concrete realities of our own classrooms. This text is designed to help mathematics teachers realize the ideal and bring the abstract to the concrete through key practices targeting the development of student ownership of learning. For when asked the question *Whose math is it?* every student should respond, *My math!*

The Role of the Students

Think about the students in your classroom. How do they see themselves as participants in the mathematics classroom community? Further, how do they see themselves in respect to math itself? Some students consider themselves to be passive recipients in the mathematics classroom—why is this? Math, to them, is likely a large collection of facts and procedures that need to be unveiled by an expert so they can be apprenticed into recall and reproduction. In this sense, mathematics is much like tradition in that it must be passed on to survive—if all the math teachers suddenly vanished we would never know math again! (Something that would likely land with minimal tragic impact to the students described here.) These students don't have a *say* in mathematics—no one does! Mathematics just simply *is*.

Contrast the mindsets of these students with those in the classroom previously described, where students are clearly positioned as problem-solvers with agency over their learning. They have a stake in the game, they lean into challenge, and they believe progress will come with effort. To those with agency in the subject, mathematics is something that can be—and *needs to be*—discovered individually and collectively. The ability and authority to validate mathematical claims, check the accuracy of calculations, and determine the reasonableness of solutions lives within them—not beyond them. They may appreciate external validation, but it is not prerequisite to confident progress. To these individuals, math is *personal*, math is *owned*. These individuals cannot be told that $1 + 1 = 47$, for they have independent access to the existential structure of mathematics where this falsehood doesn't pass the smell test. Simply put, these individuals are mathematicians.

Surely we have had students arrive in our classrooms with mindsets on both ends of the spectrum outlined here—as well as in many places in between. The question for us as teachers becomes, how do we take students from wherever they are and help them develop more of the ownership required to be successful in mathematics? In order to do this, however, we need a benchmark understanding of their foundational starting point. One way to do this is by using the *Student Mathematical Ownership Itinerary* (Table I.1 and Table I.2). This tool can be used to inform you (and your students) how each learner situates themselves in the mathematics classroom and in respect to math itself. It can be used at the beginning of the school year as a pre-assessment of mathematical agency, as a formative benchmark throughout the school year to inform your instructional decision making, and at the end of the year to measure the impact of your approach.

Table I.1   Student Mathematical Ownership Itinerary

STUDENT MATHEMATICAL OWNERSHIP ITINERARY			
State the degree to which you agree with each statement below.			
1. I can use math as a tool to make sense of the world.			
[ ] Strongly Agree	[ ] Agree	[ ] Disagree	[ ] Strongly Disagree
2. Math is a large collection of facts and procedures that need to be memorized.			
[ ] Strongly Agree	[ ] Agree	[ ] Disagree	[ ] Strongly Disagree
3. I can discover math on my own.			
[ ] Strongly Agree	[ ] Agree	[ ] Disagree	[ ] Strongly Disagree
4. I need a teacher to show me how to do math before I can learn it.			
[ ] Strongly Agree	[ ] Agree	[ ] Disagree	[ ] Strongly Disagree
5. I can make choices when doing math about how I want to solve a problem.			
[ ] Strongly Agree	[ ] Agree	[ ] Disagree	[ ] Strongly Disagree
6. There is one right way to do math.			
[ ] Strongly Agree	[ ] Agree	[ ] Disagree	[ ] Strongly Disagree
7. I can check my own work to see if I did it right.			
[ ] Strongly Agree	[ ] Agree	[ ] Disagree	[ ] Strongly Disagree
8. I need a teacher to tell me if my answers are right.			
[ ] Strongly Agree	[ ] Agree	[ ] Disagree	[ ] Strongly Disagree

online resources   Available for download at **https://companion.corwin.com/courses/whosemathisit**

Table I.2   Student Mathematical Ownership Itinerary (version 2)

STUDENT MATHEMATICAL OWNERSHIP ITINERARY

Read each statement. Circle the picture that matches how you feel.

1. I can use math as a tool to make sense of things around me.

Strongly Agree	Agree	Disagree	Strongly Disagree

2. Math is a group of facts and steps to take that I need to memorize.

Strongly Agree	Agree	Disagree	Strongly Disagree

3. I can figure out math on my own.

Strongly Agree	Agree	Disagree	Strongly Disagree

4. I need a teacher to show me how to do math before I can learn it.

Strongly Agree	Agree	Disagree	Strongly Disagree

5. I can make choices when doing math about how I want to solve a problem.

Strongly Agree	Agree	Disagree	Strongly Disagree

6. There is only one right way to do math.

Strongly Agree	Agree	Disagree	Strongly Disagree

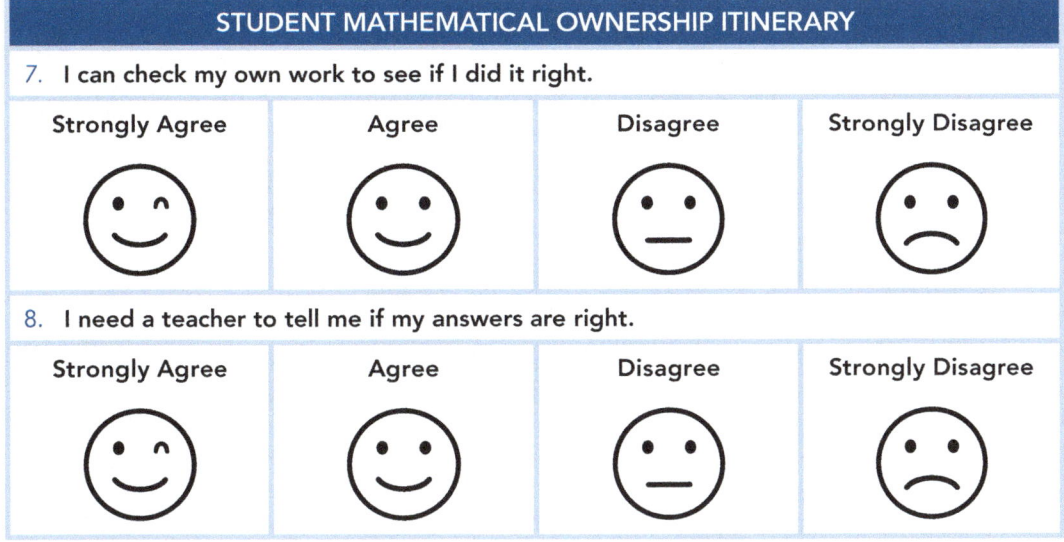

STUDENT MATHEMATICAL OWNERSHIP ITINERARY

7. I can check my own work to see if I did it right.

| Strongly Agree | Agree | Disagree | Strongly Disagree |

8. I need a teacher to tell me if my answers are right.

| Strongly Agree | Agree | Disagree | Strongly Disagree |

Source: Smiley icons courtesy of iStock.com/Makrushka

 Available for download at **https://companion.corwin.com/courses/whosemathisit**

To score this assessment, assign a scoring scale of 3: Strongly Agree, 2: Agree, 1: Disagree, and 0: Strongly Disagree to all odd numbered statements and a reversed scale of 0: Strongly Agree, 1: Agree, 2: Disagree, and 3: Strongly Disagree to all even numbered statements. Scores of 0–10 indicate low perceived ownership of mathematics, 11–16 indicate a moderate ownership of mathematics, 17–24 indicates a high level of student ownership of mathematics.

To be clear, I am not trying to send the message that students arrive in some sort of a fixed manner regarding mathematical ownership whereby some have it and some simply do not. Rather, this initial focus on the role of the student is meant to highlight the impact of their surroundings and learning environments—including their teacher—on their presumed capacity for mathematical ownership. In other words, as teachers, we have great influence over how students position themselves with mathematics. The language we use, the environments we foster, the tasks we launch, the ways we interact with others—all of this impacts how students are positioned in the content and our classroom/math course. That's great news! It means we have the power to affect positive change in our students' sense of self. If, that is, we act with intention. In the next section, I will seek to further illustrate how our decisions and actions as teachers produce much more than just marks on papers.

The Role of the Teacher

Think about our primary role as teachers of mathematics. Are we disciples of the subject, facilitators of learning, or perhaps, both? Consider the following exchange between a student and teacher during a middle school lesson on using variables to represent quantities in a real-world problem. Students are independently working on the following problem while the teacher circulates the room.

The perimeter of a rectangular swimming pool is 54 meters. The length of the pool is 6 meters. What is its width?

Student: *[raises their hand and signals the teacher over] Is this right?*

Teacher: *Can you tell me what you did?*

Student: *OK. Well, I wrote 6 · w = 54 because the formula is l · w and then just divided 54 by 6 and got 9 for w.*

Teacher: *So, that's the formula for area . . .*

Student: *Ohhh . . .*

Teacher: *. . . and you want perimeter instead, which is 2l + 2w = 54. So since you know the length is 6, you can write [signals to student to start writing as he speaks] 2(6) + 2w = 54. Right. Now what is 2 · 6?*

Student: *12?*

Teacher: *Right. And now we need to subtract the 12 from both sides of the equation [points to paper to indicate the student should write what he is suggesting]. And 54 − 12 is . . . ?*

Student: *42?*

Teacher: *OK so if 2w = 42, then how much is just one w?*

Student: *21?*

Teacher: *That's right! Make sure you write that all down. [Continues circulating room]*

What do we notice about how the teacher and student respond to one another? The student—for one reason or another—was looking for some sort of validation of their work. Work, it is worth mentioning, that was absolutely mathematically correct, albeit misplaced on this particular task. The teacher follows the student's inquiry with an open request for explanation, which could communicate the importance of process in the class. Once the student unveils their thinking, however, the teacher assumes a corrective stance and begins walking the student through the problem-solving process. The student seems to recognize their error in problem setup after the teacher informs them that "that's the formula for area," but is quickly cut off as the teacher proceeds to plow the *correct* solution path.

Let's think about what we can infer about their presumed roles and positions within that classroom. It is difficult to discern exactly how the student might presume their own role in the classroom based on this exchange, because frankly, we don't hear much from them. The teacher, however, appears to have assumed the role of Corrector-in-Chief. Which is an important and fitting role if our primary task as math teachers is to help students produce correct answers. It is clear that the teacher has situated himself as the arbiter of truth in this exchange—the master codex against which other participants might calibrate their own efforts. Now, we should be careful here not to completely demonize the familiar "sage on the stage" metaphor—for content expertise is an invaluable tool to facilitate the many roles teachers must navigate to promote a student-centered classroom. However, the consideration I am promoting here is regarding the impact the teacher is having on the student's sense of ownership in the content and classroom/course. Namely, how is the teacher's own positioning as the *knower* and *shower* affecting that of the student? Well, we can only infer based on what we see. The student was situated to only follow instructions and answer tightly close-ended calculations. Here are some reasonable conclusions from this exchange:

- *The teacher sets up the problem, and I solve it.*
- *I need to do this like the teacher.*
- *Calculations are the important part.*
- *The way I did it was wrong.*

Regardless, are these the messages that foster student ownership in mathematics? How might this student respond if we asked them *whose math is it?*

There was a clear decision-point for the teacher in this exchange after the student explained their thinking. Let's take a look at the same exchange again, this time highlighting the decision-point, along with some additional considerations on the part of the teacher and alternative responses. We will use the expert noticing framework (Jacobs et al., 2010) whereby we first attend to the details of the case, then interpret their meaning, and finally choose how to respond.

The perimeter of a rectangular swimming pool is 54 meters. The length of the pool is 6 meters. What is its width?

Student: *[raises their hand and signals the teacher over] Is this right?*

Teacher: *Can you tell me what you did?*

Student: *OK. Well, I wrote 6 · w = 54 because the formula is l · w and then just divided 54 by 6 and got 9 for w.*

Decision-Point

Expert Noticing: This student is correctly using variables to represent unknown quantities and is correctly solving for those quantities. However, this student set up the problem as if they were given the area of the pool of 54 square meters rather than the perimeter of 54 (linear) meters. There is a possibility that there is confusion around units (meters versus square meters), but it could have just been an oversight, and that also isn't the primary focus of this task. There is also a possibility that the student does not know the difference between area and perimeter, but that is not clear yet, so I will need to gauge more about this. Also, I want to be careful to honor the work the student has done and situate it as legitimate mathematics, though different than what the task is seeking. So, I want to use language that validates *their* process.

Teacher: *OK. I see what you did here, and I appreciate how you used variables to represent the unknown quantities. I heard how you talked through your problem-solving process and calculations, and it all sounded mathematically legitimate to me. So here's my question . . . How would you do this if the AREA of the pool was 54 square meters instead?*

Student: *[Silent for a moment while looking at their work, and the original problem.] The area is 54? Oh, OHH!!!*

Teacher: *Yup, there it is.*

Student: *Ahhh I did area instead of perimeter! [Starts erasing]*

Teacher: *Yeah you did and . . . [waves hands] No, no! Don't erase it! That's really great work for a different problem. Maybe we should even give it to the class next? Just write the new work for this problem underneath.*

Now what do we notice about how the teacher and student respond to one another? And what can we infer about their presumed roles and positions within that classroom? In contrast to the first exchange, this time the teacher led with validation and recognition of the student's legitimate mathematical thinking—which was not contrived. Then, we saw the teacher guide the student's thinking with a targeted question that held multifaceted value. Asking the student about area provided the teacher insight into whether the student recognized the difference between area and perimeter (one of the early content wonderings), as well as served as a prompt to trigger the student's thinking around the actual *ask* of the task. The teacher did not jump into premature reteaching—which would have served as a rigor-reducing overscaffold in this case.

Further, the teacher communicated confidence in the student's own recognition of what adjustments needed to be made, which could reinforce the student's sense of ownership and efficacy in mathematics. Finally, the teacher made very clear that the student should not *undo* their original work by erasing it. This final validating move of the student's thinking could only continue to perpetuate the message that their contributions matter and their mathematical thinking is worthy. So then, perhaps some reasonable conclusions from this second exchange might include the following:

- *The teacher is here to guide me but not do the work for me.*
- *Sometimes I need the teacher, and sometimes I don't.*
- *Calculations are important but so is correctly setting up a problem.*
- *The way I did it was right but for a different problem.*

Regardless, these contrasting messages could serve to foster greater student ownership in mathematics. How might this student now respond if we asked, *whose math is it?*

Our decisions in the classroom, our choices during planning, and the way we respond to students all have the propensity to greatly affect how students see themselves as mathematicians. We have the power to contribute to or detract from our students' sense of agency and mathematical ownership—all of which contributes to their ever-dynamic identities. We need to act with care, and we need to act with intention if we are to use our powers for good. Thus is the intent of this book. How can we structure our courses, classrooms, and ourselves toward this end of promoting mathematical ownership in our students?

How to Use This Book

This book is rooted in teacher clarity and split into two parts, both presented through the context of mathematics education: determining success criteria and operationalizing success criteria. The first part, Determining Success Criteria, is intended to help teachers clearly define success in mathematics in a way that is productive for their students. We will also look at relevant research and best practices, which is the focus of Chapter 1. The second part, Operationalizing Success Criteria, is intended to help teachers provide opportunities for students to build their success and ownership in mathematics in whole-class, peer-to-peer, and individual settings through the development of social and sociomathematical norms, collaborative learning experiences, and self-regulated learning.

Sociomathematical norms: norms that are specific to a *mathematics* learning community and regulate the community's communication about and participation with the subject of *mathematics*.

Chapter 2 will explore the teacher's role in developing classwide social and sociomathematical norms that underpin the mathematical culture of their classrooms. It will also discuss how to leverage the clarity gained in Chapter 1 to explicitly develop, maintain, and leverage social norms with social learning intentions. We will see that sociomathematical norms develop in any learning community whether we intend them to or not, for better or worse—so we ought to consider shaping them with intention. Chapter 2 will further illustrate how to communicate and model the existence of *choice* in mathematics, as well as how to use discursive positioning moves to situate our students as problem-solvers with agency. The mantra for mathematical ownership at the whole-class level in this chapter is *everybody's doing it*.

Chapter 3 will discuss how to reinforce student ownership by structuring peer interactions and collaboration and will make the case for investing in collaboration as a space for students to begin taking ownership of their learning. Importantly, this chapter will recognize that students need to be primed in order to ensure that group work is indeed productive. Everything from grouping strategies to setting up and launching tasks will be covered to this end. This chapter also serves as a hub for various collaborative strategies and protocols suitable for the mathematics classroom. The mantra for mathematical ownership among students at the peer-to-peer level in this chapter is *we're doing it*.

Chapter 4 homes in on supporting individual students by promoting metacognition and self-regulated learning—essential components of ownership. It will delineate the self-directive process of self-regulation into its individual components and discuss how to scaffold students toward increased motivation by targeting each for development. This includes teaching students how to become more independent learners and study. Finally, it will demonstrate the importance of feedback and student self-assessment in self-regulated learning. The mantra for mathematical ownership for students individually in this chapter is *I'm doing it*.

The book closes with a review of the student-facing mantras of this book and their implications, as well as provides some teacher-facing mantras to guide classroom policies and decision making. Implementation is as much about mindset as it is about action. Building student ownership of mathematics requires both a plan *and* a sense of direction. I aim to ensure this book provides both. The intent of this closing section is to facilitate a sense of ownership in the reader and communicate that *You can do it*.

Each chapter will begin with its own overarching learning intention and set of specific success criteria to ground your learning by communicating our goals. Success criteria will have additional callouts throughout each chapter to model *signaling*, an aspect of teacher clarity discussed in the next chapter that helps guide learning by providing additional structure. Each chapter will conclude with reflection questions, to help you make personal connections to your own practices and mathematical experiences, as teacher clarity also encompasses understanding ourselves. Speaking of clarity—let's start there.

PART I

Determining Success Criteria

What Does It Mean to Be Successful in Mathematics?

1

Teacher clarity is more than a lesson plan; it's a sense of direction. Clarity allows instruction to be intentional and learning to be purposeful. After all, *"every student deserves a great teacher, not by chance but by design"* (Fisher et al., 2016). Pursuing clarity is the act of intentionally designing the great teacher that your students deserve. The argument for clarity, then, is simple: How can one expect to achieve any sort of outcome if that intended outcome is unknown? In other words, how do we expect to hit a target we aren't aiming for? If we are to make the largest learning gains with our students, and promote ownership of their learning and of the content itself, then we ought to begin by spending some time clarifying what it means to be successful in mathematics. What is it, exactly, that we should want for our students?

Teacher clarity is more than a lesson plan; it's a sense of direction

This chapter is dedicated to this concept of teacher clarity but specifically tailored for mathematics. I will begin by further defining teacher clarity and building the case for its pursuit—which is hard work! From there we will build consensus around what is meant by *success in mathematics* by compiling and examining existing works in the mathematics education community.

CHAPTER 1

Learning Intention:

I am pursuing teacher clarity by learning what it means to be successful in mathematics.

Success Criteria:

☐ I can explain the value of teacher clarity and the positive impact it has on students.

☐ I can identify connections between the Five Strands of Mathematical Proficiency, the Standards for Mathematical Practice, and the three aspects of mathematical rigor.

☐ I can define *success* in mathematics.

The Importance of Clarity and Measuring Success

Teacher clarity:
a measure of clarity of communication between teachers and students in both directions.

Teacher clarity is not a new concept. It has been studied and measured for the last half century in a variety of K–12 and college instructional settings. Fendick (1990) defined teacher clarity as "a measure of the clarity of communication between teachers and students in both directions" (p. 10). Fendick (1990) conducted a meta-analysis investigating teacher clarity across four dimensions, illustrated in Figure 1.1. The four dimensions that constitute his definition are (1) Clarity of Organization; (2) Clarity of Explanation; (3) Clarity of Examples and Guided Practice; and (4) Clarity of Assessment of Student Learning. Let's consider each of these.

Clarity of organization occurs at the lesson, the unit, and the whole year/whole-course level and includes such features as determining and stating learning intentions, aligning the content to formative and summative assessments, and reviewing content throughout the year/course.

Clarity of examples and guided practice refers to keeping instruction aligned to assessments, interacting formatively with students, providing time for practice, providing metrics of success (success criteria), and providing students with formative feedback (Fendick, 1990).

Figure 1.1   Four Dimensions of Teacher Clarity

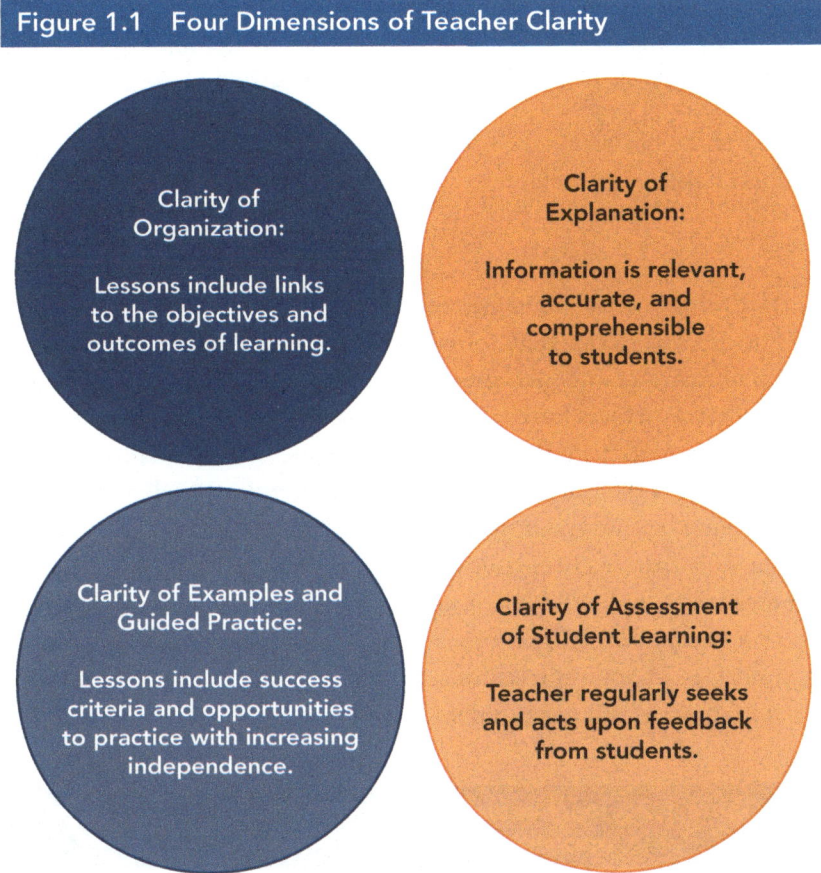

Clarity of explanation is how the teacher simplifies or clarifies explanations and infuses them with relevance. It also refers to how a teacher emphasizes and reemphasizes directions and key points, connects content to prior knowledge, and pursues appropriate pacing based on student understanding and concept mastery.

Clarity of assessment of student learning includes the methods the teacher uses to check for understanding throughout a lesson, encourage class discussion, and provide feedback on assignments and assessments (Fendick, 1990).

Everything that occurs during instruction should be intentionally linked toward some common end (what students need to learn)—and that common end should be understood by both teachers and students.

CHAPTER 1 SUCCESS CRITERIA CALLOUT:

☐ I can explain the value of teacher clarity and the positive impact it has on students.

Fendick (1990) was not the only researcher to determine that teacher clarity has a positive influence on student learning. Houser and Frymier (2009) studied how both student characteristics (namely, temperament and learner orientation) and teacher behaviors (nonverbal immediacy and clarity) influenced student empowerment. Both teacher immediacy and teacher clarity were found to have a greater effect on empowerment than any student characteristics measured in their study. Further, teacher clarity was found to have a direct impact on learning outcomes in addition to its indirect effect on outcomes through empowerment. This builds the case that teachers have the power to directly impact student empowerment, which is a precursor to efficacy. The evidence suggests that teachers have a greater impact on students' empowerment than students do themselves—which arguably reframes this task as a *responsibility* for teachers. Figure 1.2 demonstrates how teacher clarity has both a direct impact on student learning and an indirect impact on learning via student empowerment.

Figure 1.2 Teacher clarity has both a *direct* and *indirect* impact on student learning.

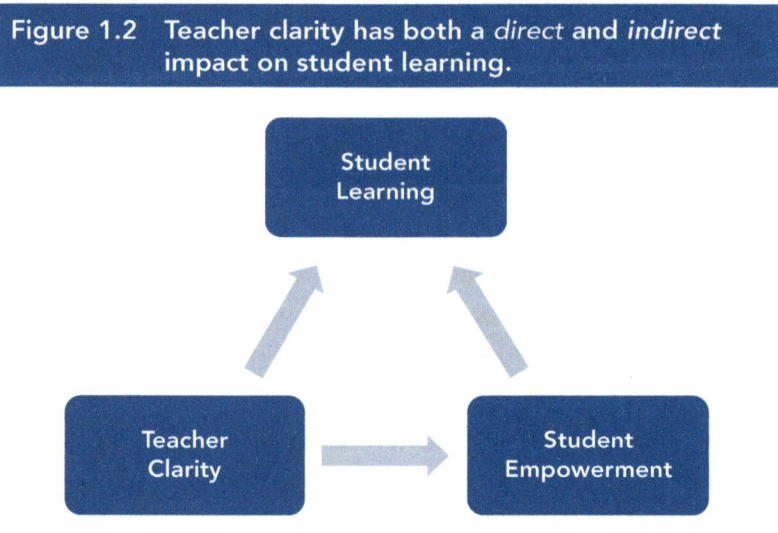

Further building this case, Titsworth et al. (2015) conducted two meta-analyses supporting the positive effects of teacher clarity on student learning. The results showed affective learning was impacted more than cognitive learning (200% increase versus

100% increase). Figure 1.3 further compares and contrasts the affective and cognitive domains of learning. This emphasis on affective learning would appear to align to Houser and Frymier's (2009) findings on teacher clarity's impact on student empowerment. Increases in teacher clarity also reduce the cognitive load of learning and increase motivation (Serki & Bolkan, 2024).

Signaling is another specific aspect of teacher clarity, which involves teaching with instructional and organizational cues that make the structure of the lesson transparent to students (Bolkan, 2017). In practice, this can commonly be seen as the interweaving and revisiting of learning intentions and success criteria throughout instruction. (You may have noticed this book making use of signaling with the Chapter Success Criteria Callouts.) For many students, signaling is an organizational and structural scaffold that frees up working memory for other tasks, such as learning the actual mathematics content of the lesson. These findings arguably situate teacher clarity as an equitable teaching issue.

Figure 1.3   Teacher Clarity's Impact on Student Empowerment

Cognitive Domain vs. Affective Domain

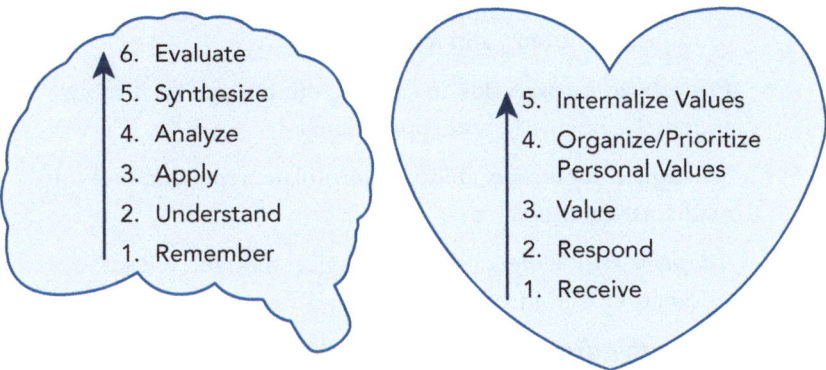

The understanding of content knowledge that develops from basic to complex as learners engage in tasks (Bloom et al., 1956)

The dispositions, emotions, attitudes, and feelings learners experience while engaging in tasks, as well as how these develop as the tasks progress in complexity (Bloom et al., 1964)

Source: Adapted from Bloom, et al., 1956 and Bloom, et al., 1964. Brain outline courtesy of iStock.com/Anatoliy Stepura

With the case being made for the benefits of teacher clarity, how might we actually begin implementing this measurable influence on learning? We should set our eyes on the prize: What do we and should we *intend* for our students? What does it mean to be successful in mathematics? The five strands of mathematical proficiency are key to answering these questions.

CHAPTER 1 SUCCESS CRITERIA CALLOUT:

☐   I can identify connections between the Five Strands of Mathematical Proficiency, the Standards for Mathematical Practice, and the three aspects of mathematical rigor.

The Five Strands of Mathematical Proficiency

For some, success in school mathematics is simple to define: right answers signal success, while wrong answers signal failure. And while this perspective holds an obvious validity from a very literal point of view, research has expanded our view of success to include emphasis on both procedures and understanding. So if right and wrong answers are no longer the determining factors of success in mathematics, what measures do we replace them with? Enter the Five Strands of Mathematical Proficiency (Kilpatrick et al., 2001). The National Research Council defines *mathematical proficiency* as existing across five strands:

- *Conceptual understanding*: comprehension of mathematical concepts, operations, and relations

- *Procedural fluency*: skill in carrying out procedures flexibly, accurately, efficiently, and appropriately

- *Strategic competence*: ability to formulate, represent, and solve mathematical problems

- *Adaptive reasoning*: capacity for logical thought, reflection, explanation, and justification

- *Productive disposition*: habitual inclination to see mathematics as sensible, useful, and worthwhile, coupled with a belief in diligence of one's own efficacy (Kilpatrick et al., 2001)

This work places special emphasis on the fact that understanding is greater than memorization, that the connections garnered by deep learning are prerequisite to transferring knowledge to novel situations, and that metacognition and motivation are both pivotal to learning. Kilpatrick et al. (2001) argue that, "[m]athematical proficiency . . . cannot be achieved by focusing on just one or two of these strands," but note that it is also not a simple dichotomy of either proficient or not (p. 116). So, mathematical proficiency should be developed along each strand, across the strands, and over time. The intertwined nature of these five strands is further illustrated in Figure 1.4. The remainder of this section will be dedicated to fleshing out each of these strands.

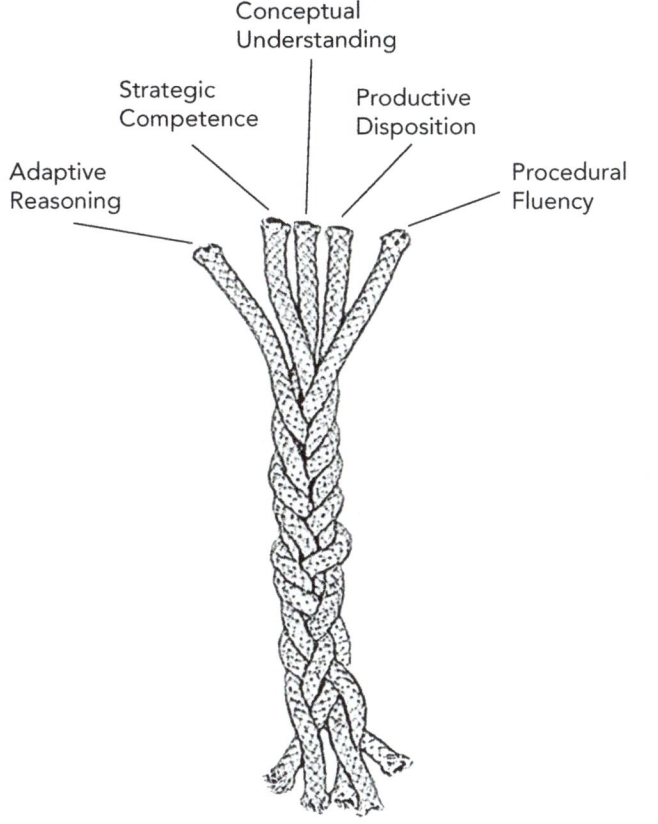

Figure 1.4   Five Intertwined Strands of Mathematical Proficiency

Source: Used with permission of The National Academies Press from *Adding it up: Helping children learn mathematics*, Kilpatrick et al., 2001; permission conveyed through Copyright Clearance Center, Inc.

Conceptual Understanding

Conceptual understanding manifests itself when students comprehend mathematical concepts, operations, and relations in a functional and integrated way (Kilpatrick et al., 2001). The connected nature of conceptual understanding organizes content and allows new knowledge to *fit*, rather than live in isolation. This has multiple benefits for students. For example, students are able to recognize new iterations of old concepts, or, "superficially unrelated situations" (different in appearance or context only), thus resulting in less to learn (Kilpatrick et al., 2001). Another benefit is that students recognize errors as they occur, largely because they don't *fit* existing expectations. This integration of knowledge also improves students' retention of information and breeds confidence by reinforcing their own sense of logical reasoning and natural conclusions. For example, when a student actually understands a concept, forgotten details can be logically reconstructed and checked against metrics of reason rather than just hoping that the information will pop back into their head somehow.

Sometimes proving challenging to measure, early conceptual understanding can exist in a learner prior to their ability to demonstrate it. Generally, however, conceptual understanding can be elicited through expressing connections between representations and concepts (via concept maps or other linking diagrams, explanations, etc.). Conceptual understanding can also manifest through a student's ability to produce multiple representations of the same mathematical situation, compare and contrast each, and on a deeper level recognize the contextual usefulness of each representation.

One common strategy for teaching and assessing conceptual understanding in mathematics is the Frayer Model (Frayer et al., 1969). Often undersold as a vocabulary instruction tool, the Frayer Model helps students flesh out the contours of concepts by considering the definition, examples of the concept, nonexamples, and a pictorial representation or other characteristics. As an assessment tool, Frayer Models can be provided without the term present in the center. Students are to read the definition, explore the characteristics, examples and nonexamples, and then try to determine which concept is being represented. Figure 1.5 shows both of these approaches.

Procedural Fluency

Procedural fluency is about possessing procedural skills and a sense of direction about when to use them, as well as a certain degree of automaticity (Kilpatrick et al., 2001). In addition to promoting students' mathematical independence, procedural fluency provides students a lens into the well-structured nature of mathematics that can be generative toward their own inquiry. In other words, a high degree of procedural fluency provides students with an ability to conduct tests and experiments within and upon mathematics.

> Procedural fluency is about possessing procedural skills and a sense of direction about when to use them, as well as a certain degree of automaticity

Figure 1.5a   Frayer Model Example 1

Definition:

The result from addition.

Image or Characteristics:

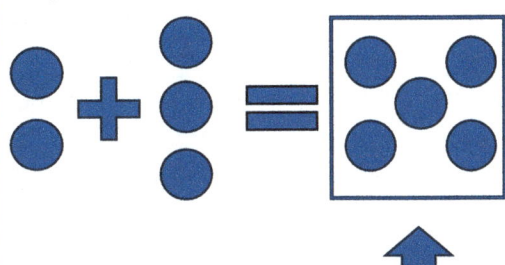

Concept:

Sum

Examples:

$3 + 2 = 5$

$7 = 5 + 2$

$$\begin{array}{r} 8\ 7 \\ +\ 2\ 1 \\ \hline 1\ 0\ 8 \end{array}$$

Non-examples:

$10 - 3 = 7$

$3 \times 4 = 12$

$100 \div 10 = 10$

Figure 1.5b   Frayer Model Example 2

Definition:

The set of numbers that includes whole numbers and their opposites.

Image or Characteristics:

Has no fractional or decimal parts

- Can be positive
- Can be negative
- Can be zero
- Can be modeled with two color tiles

Concept:

Examples:

−7	12	−9,878
1	−1	1,000,000
0	400	

Non-examples:

1.3	1/2	−1.754
π	−43	$3\frac{1}{4}$
−5/4		

Fluency also frees up working memory for students to engage in the other strands of mathematical proficiency by easing the difficult and simplifying the complex through familiarity.

The National Council of Teachers of Mathematics (2014) argues through their Mathematics Teaching Practices that procedural fluency must be built on a foundation of conceptual understanding. Further, teaching isolated procedures first can make students resistant to investigating their conceptual underpinnings (Kilpatrick et al., 2001). Also, rooting procedures in conceptual understanding is more efficient and will require less commitment to memory and less practice toward retention, which again speaks to the connected nature of conceptual understanding (Kilpatrick et al., 2001). As teachers, this means that we should not choose *between* teaching concepts or procedures but instead should thoughtfully sequence them.

Strategic Competence

Strategic competence goes beyond problem solving to include two of its precursors, formulating mathematical problems and representing them (Kilpatrick et al., 2001). This strand encompasses much of what would be considered mathematical modeling, as well as the related process of mathematizing the real world or realistically imaginable (Gravemeijer & Doorman, 1999) and speaks to the field of applied mathematics. While mathematics is indeed the universal language, it is rarely spelled out for us in the real world. This is why students need to be able to understand situations and their key features and then mathematize the relevant features while ignoring those that are irrelevant to the problem or situation (Kilpatrick et al., 2001). This mathematical modeling process could involve crafting an equation or some other representation of the situation, such as a drawing or diagram. For instance, elementary students might be tasked with calculating the area of sports fields or other grassy areas by mathematizing an aerial view of their actual shapes into composites of known two-dimensional shapes. Secondary students might be tasked with measuring the heights of buildings and other tall objects using an inclinometer and trigonometry. Broadening students' exposure from simply routine problems to also include non-routine problems such as these that require productive and inventive thinking builds flexibility with strategic competence. Students develop further fluency through strategic competence when they translate their hindsight gained from a nonroutine problem into foresight the next time they engage in a similar task.

Adaptive Reasoning

Adaptive reasoning is possibly the most challenging strand to unwind from the others, as its impact is absolutely pervasive throughout all mathematics. Adaptive reasoning speaks to the logical connections between mathematical ideas and one's ability to communicate those connections. This strand is from where all mathematicians gain their authority by borrowing it from the purity of the subject itself. Adaptive reasoning empowers students to be their own metric of success by tapping into their own sense of reason. Adaptive reasoning is a broad idea that encompasses both formal and informal modes of communication, such as proofs, oral discussion, informal explanation and justification, and intuitive reasoning based on patterns, analogies, metaphors, and other thinking tools. Thus, adaptive reasoning is measurable through students' ability to justify their work and explain their thinking. And while any success criteria aimed at measuring adaptive reasoning should absolutely be operationalized at the individual level, this area also begs the use of social and sociomathematical norms, which will be discussed in the next chapter.

Productive Disposition

Productive disposition is about learners recognizing the value of and thus caring about mathematics, while also believing that they have the capacity to learn it. Students who have a productive disposition toward mathematics believe that the subject is understandable, it should make sense, and, with appropriate effort, is conquerable and worthwhile. Most children begin their schooling with a productive disposition toward mathematics; however, this often changes for many somewhere along their academic careers depending on their experiences (Kilpatrick et al., 2001). Further, students' experiences shape their beliefs about mathematics and hence their dispositions toward the subject. This situates teachers in a very powerful and consequential position to shape *and maintain* students' productive dispositions toward mathematics.

> Productive disposition is about learners recognizing the value of and thus caring about mathematics, while also believing that they have the capacity to learn it.

The Standards for Mathematical Practice

While most standards documents tend to delineate the various concepts and procedures to be gained through the content standards of each course, many have also begun fleshing out additional practice standards that embody the latter three strands of mathematical proficiency. Practice standards are a wonderful addition for this reason. They provide a space to explicitly call out the habits of mind, dispositions, and approaches identified in the Strands of Mathematical Proficiency that would otherwise only be implicit (at best) in content standards. These practice standards should not be viewed as additive but rather as clarifying for our task as mathematics teachers. Consider the alignment in Table 1.1 between the eight Standards for Mathematical Practice (National Governors Association Center for Best Practice, Council of Chief State School Officers, 2010) and the five Strands of Mathematical Proficiency.

Table 1.1   Alignment Between the Standards for Mathematical Practice and Five Strands of Mathematical Proficiency

STANDARD FOR MATHEMATICAL PRACTICE	RELATED STRANDS OF MATHEMATICAL PROFICIENCY
1. Make sense of problems and persevere in solving them.	Productive Disposition
2. Reason abstractly and quantitatively.	Adaptive Reasoning; Strategic Competence
3. Construct viable arguments and critique the reasoning of others.	Adaptive Reasoning
4. Model with mathematics.	Strategic Competence
5. Use appropriate tools strategically.	Strategic Competence
6. Attend to precision.	Adaptive Reasoning
7. Look for and make use of structure.	Adaptive Reasoning; Strategic Competence
8. Look for an express regularity in repeated reasoning.	Adaptive Reasoning; Strategic Competence

Rigor: the balanced pursuit of conceptual understanding, procedural skills and fluency, and application with equal intensity

Three Aspects of Rigor in Mathematics

Rigor is the balanced pursuit of conceptual understanding, procedural skills and fluency, and application with equal intensity (Common Core State Standards Initiative [CCSSI], 2020). To more fully understand how rigor is reflected in content standards, let's look at each aspect of the definition.

- **Conceptual understanding:** The standards call for conceptual understanding of key concepts, such as place value and ratios.

Students must be able to access concepts from a number of perspectives in order to see math as more than a set of mnemonics or discrete procedures.

- **Procedural skills and fluency:** The standards call for speed and accuracy in calculation. Students must practice core functions, such as single-digit multiplication, in order to have access to more complex concepts and procedures. Fluency must be addressed in the classroom or through supporting materials, as some students might require more practice than others.

- **Application:** The standards call for students to use math in situations that require mathematical knowledge. Correctly applying mathematical knowledge depends on students having a solid conceptual understanding and procedural fluency.

In order to have clarity around instructional design and sequencing, it is important to recognize and understand these three aspects of rigor both independently and interdependently. Some standards, for instance, call for each of these aspects of rigor in isolation. For example, consider the following second-grade standard regarding number and operations in base ten:

> 2.NBT.1 Understand that the three digits of a three-digit number represent amounts of hundreds, tens, and ones; e.g., 706 equals 7 hundreds, 0 tens, and 6 ones. Understand the following as special cases: a. 100 can be thought of as a bundle of ten tens—called a "hundred." b. The numbers 100, 200, 300, 400, 500, 600, 700, 800, 900 refer to one, two, three, four, five, six, seven, eight, or nine hundreds (and 0 tens and 0 ones). (Taken from CA Mathematics standards)

By use of the verb *understand*, it is clear that this standard is specifically targeting a student's conceptual understanding of place value. No action other than understanding is called out and thus expected of students. Teachers would do well to recognize this and plan instruction with this target in mind.

Similarly, consider this sixth-grade standard regarding number sense and calculations with decimals:

> 6.NS.3 Fluently add, subtract, multiply, and divide multi-digit decimals using the standard algorithm for each operation. (Taken from CA Mathematics standards)

By similar investigation of the verbs in the standard and actions expected of students—namely to *fluently add, subtract, multiply, and divide*, as well as to *use the standard algorithm*, it is clear that this standard is aiming to build students' procedural skills and fluency. This recognition would serve teachers well when identifying the learning intentions and determining what success would look like for this standard.

Finally, consider this high school integrated Math II standard regarding geometric measurement and dimension:

> G-GMD.6 Verify experimentally that in a triangle, angles opposite longer sides are larger, sides opposite larger angles are longer, and the sum of any two side lengths is greater than the remaining side length; apply these relationships to solve real-world and mathematical problems. (Taken from CA Mathematics standards)

The use of the term "verify experimentally" signals that students should be engaging in hands-on experiences, and "real-world" makes it clear that the goal of this standard is for students to apply their mathematics to novel situations.

Notably, this final standard is asking students to apply their procedural skills and conceptual understanding around solving side and angle measurements for triangles. This should not be understated as many standards documents state that, "Correctly applying mathematical knowledge depends on students having a solid conceptual understanding and procedural fluency" (CCSSI, 2020). To state this plainly, conceptual understanding is a prerequisite to procedural skills and fluency, and both are prerequisite to application. While application is indeed the leveraging of both conceptual understanding and procedural skills, this does not isolate it to the final phases of instruction. Application can be used throughout a unit of study to discover the need for new procedures or to provide access to a concept through real-world familiarity (viral videos, purchasing things from a menu, etc.). It also exists as the ultimate display of mathematical ownership of given concepts and skills. Figure 1.6 illustrates this balanced approach to mathematics teaching and learning.

In stark contrast to this balanced approach, the traditional math classrooms that many of us experienced as students ourselves were almost entirely focused on procedures (see Figure 1.7). The idea here was to arm us all with the tools we would use later in calculus. (How is *that* for

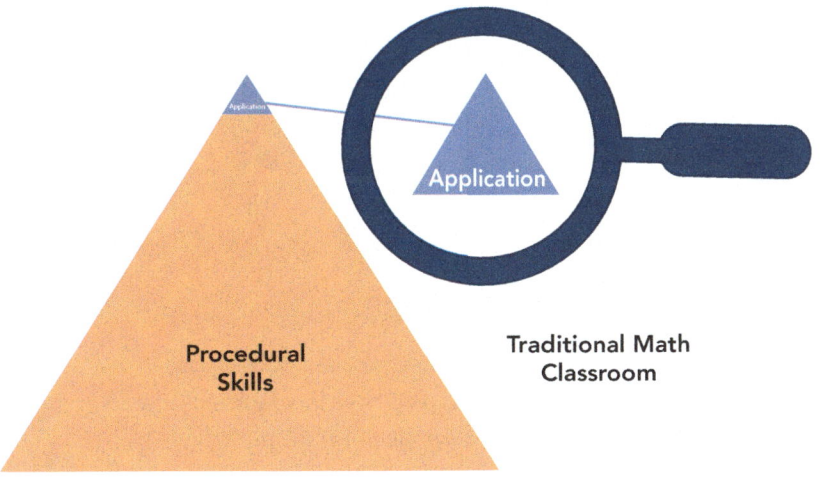

Figure 1.6   A Balanced Approach to Mathematics Instruction

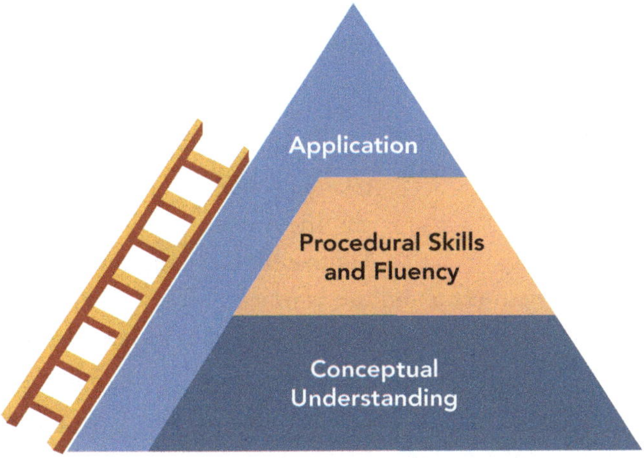

Figure 1.7   An Unbalanced Approach to Mathematics Instruction in a Traditional Math Classroom

a faith-based approach to relevance? *Learn this—I promise you'll need it later!*) Application was saved for the end of each section or chapter *if we had time*—which meant it was often cut. An important reckoning for some of us, however, is that we accidentally developed a conceptual understanding along the way. As teachers of math now ourselves, we must be careful not to give more credit than is due to the system that built us. We didn't learn math and develop a sense of ownership *because* of the traditional approach; we did so *despite* it.

Determining Our Success Criteria

So, how is it that a student develops a productive disposition, conceptual understanding, procedural fluency, and leverages those toward strategic competence, adaptive reasoning, and application—all the while exhibiting the habits of mind and practices of a mathematician? This task requires students to take *ownership* of the content and situate themselves as active practitioners. Ultimately, then, success in mathematics is the propensity to learn and use as much of it as one needs or wants. This is mathematical ownership. When asked *Whose math is it?* successful students will respond *my math!*

CHAPTER 1 SUCCESS CRITERIA CALLOUT:

☐   I can define *success* in mathematics.

If this is what it means for students to be successful in mathematics, then our task as mathematics educators is to create spaces that promote productive dispositions and set the stage for strategic competence, to interact in ways that develop agency, and to create experiences that foster increasing independence and adaptive reasoning. This is the vehicle and setting through which we will be able to teach toward a balance of conceptual understanding, procedural fluency, and application. *Clarity achieved.* Now that we have a goal in mind we can begin taking steps toward reaching it. The second part of this book will seek to do exactly that.

CHAPTER 1 REFLECTION QUESTIONS

- What are some examples from your classroom of students' productive dispositions changing over the course of a year? What might the catalysts have been for these changes?

- What does the balance of conceptual understanding, procedural skills and fluency, and application look like in your classroom? How does this compare to your own experience as a math student?

- As you consider the Strands of Mathematical Proficiency, where are you finding opportunities to build student ownership across each?

PART II

Operationalizing Success Criteria

Developing Classwide Social and Sociomathematical Norms

2

Everybody's Doing It

A misnomer exists in education that classroom culture is something that can be *set*—as if the construct of culture is a box to be checked at the onset of the school year. (Imagine if it were that easy?) In reality, classroom culture is something that develops over the course of time in a community—with or without direct input from the teacher. The messages that are sent through the ways in which the teacher interacts with individual students, the class as a whole, and other colleagues all serve to shape classroom culture. Further, the way students interact with one another, as well as how the teacher responds to those interactions, also serve to define classroom culture. Classroom culture is the result of the ever-present and ongoing feedback loop between community members. It is dynamic and organic, but it can absolutely be (and *should* be) molded by the teacher so that it may be leveraged toward success in mathematics. As teachers, we hold the greatest influence in shaping classroom culture, so let's use that influence to shape a culture that promotes student ownership of mathematics.

This chapter is dedicated to understanding how culture develops in the mathematics classroom and the role of the teacher in that development. Over the next few sections, I will delineate classroom culture into some of its important constructs so that we can target each for productive shaping, maintaining, and leveraging. I will also discuss a number of tools that teachers can use to take a proactive and directive role in creating and sustaining a classroom culture that promotes student ownership of mathematics. I aim to communicate that while classroom culture is incredibly dynamic and complex, shaping it toward success in mathematics ultimately starts with us. When we act with intention, we can situate students as productive contributors in a mathematical community of practice and thus position them in an ownership role.

CHAPTER 2

Learning Intention:

I am learning how to develop, maintain, and leverage a classroom culture that promotes student ownership of mathematics.

Success Criteria:

☐ I can describe the impact of both social norms and sociomathematical norms in developing mathematical ownership in my students.

☐ I can use social learning intentions as a vehicle to develop and leverage social norms in my classroom.

☐ I can take a proactive approach to shaping productive sociomathematical norms in my classroom.

☐ I can promote choice in mathematics to build student ownership.

☐ I can use discursive positioning moves to situate all students as productive contributors during whole-class discussions.

The Components of a Mathematics Classroom Culture

School mathematics occurs in the social context of a classroom community. While potentially obvious, this reality is not something that should be brushed over—as communities develop culture and culture is consequential. One of these consequences of culture is the development of norms, or, communally accepted ways of being. Norms are truths of process that come from experience. Many readers are likely familiar with the concept of *classroom norms*, which exist at the functional level and consist of routines and procedures. Examples include how classroom materials are managed, expectations around how class starts and ends, caring for the classroom environment, and many others. These will not be the focus of this chapter—which is not to diminish their necessity! While classroom norms have their purpose, procedures around scissors and glue sticks are not likely to have a large impact on a student's sense of mathematical wherewithal. Rather, we will examine the highly consequential social and sociomathematical norms that are negotiated and renegotiated throughout a school year and underpin the mathematical culture of our classrooms.

CHAPTER 2 SUCCESS CRITERIA CALLOUT:

☐ I can describe the impact of both social norms and sociomathematical norms in developing mathematical ownership in my students.

I believe mathematics to be a personal dish best served socially—admittedly a view rooted in social constructivism.

From this lens, individual learning results in the development of a personal belief that is negotiated between the individual and their learning community. When an individual's personal understanding of a concept and their social interactions around that concept meet, learning has occurred (Cobb & Yackel, 1996; Rasmussen et al., 2003; Yackel & Rasmussen, 2002). This articulation process between the individual and community is essential to learning and has three major byproducts:

- **Social norms:** regularities in social interactions within a community, such as what constitutes a complete response

- **Sociomathematical norms:** the mathematical criteria for social norms having been met, such as an answer in context with appropriate units and an explanation of all procedures used with justification for each

- **Classroom mathematical practices:** mathematical actions that become "taken as given" from their frequent use, such as using the area formula after it has been developed and proven using unit cubes, or using the quadratic formula as a legitimate means of finding roots after it has been developed and proven from the standard form of a quadratic function (Cobb & Yackel, 1996, pp. 175–180)

Each of these three constructs have enormous implications as to how student ownership and mathematical autonomy are developed and will thus be further delineated in this section (Yackel & Cobb, 1996).

Social Constructivism: a perspective on learning that theorizes individual learning results in the development of a personal belief that is negotiated between the individual and their learning community.

Social norm:
a regularity in social interactions within a community that has been negotiated by the members of the community.

Social Norms

A social norm is a regularity in social interactions within a community that has been negotiated by the members of that community. A member of a community might describe their community's social norms as, *how we do things*. Social norms define how communities address certain types of occurrences, how problems are approached, how discussions play out, turn-taking structures, expectations amongst and between community members, and generally how members interact and what roles they take (Cobb & Yackel, 1996; Rasmussen et al., 2003; Yackel & Rasmussen, 2002). Through this, social norms can begin to define the very subject being addressed by the community—though the norms themselves are not subject specific. Do community members (students) view themselves as active participants in a field of study or passive spectators? The answers to questions like these can be largely impacted by the social norms of the learning community.

Social norms are negotiated and renegotiated by a community throughout the course of its existence. For example, the students in a given class period of a year-long eleventh grade math course constitute a learning community as do the students in Mrs. Hickman's third-grade class. Throughout the school year, these learning communities will renegotiate and modify social norms that were initially negotiated at the onset of the year/course. If shaped properly, these social norms can contribute to meaningful learning. If shaped improperly, these same norms can detract from community members developing a meaningful understanding of mathematics. Further, these community-wide patterns of interaction can shape how an individual views themself in regard to mathematics *and* how an individual views mathematics itself. These views, or *beliefs*, held by the individual are highly consequential to the development of a meaningful understanding of content (Yackel & Rasmussen, 2002). In this same vein, social norms can greatly impact a student's agency, which could then lead to a shift in beliefs or identity. Likewise, a student's beliefs and identity can impact which roles (defined by social norms) they assume in the community. These same beliefs also come back to bear during the negotiation and renegotiation of norms. As can be seen, these two concepts are extraordinarily intertwined.

Imagine, for instance, a middle school classroom where the teacher often presents questions to the students in an open, whole-class forum. Further, imagine that only a handful of high-achieving students respond to these questions regularly, they are often correct in their first response, and the teacher accepts their responses without eliciting responses from all students. The regularity of these happenings could promote the social norm that only some specific students share in class and others do not. It is also feasible that students could attach additional meaning to this norm that *only the kids who are really good at math share in class because they get right answers fast*—a norm and belief reinforced by the teacher's actions. This social norm could reinforce reflections by some students that, *I am not someone who shares in class because I don't get right answers fast* (a belief), *so I am not really good at math* (an indicator of identity). It is important to recognize that regular interactions and occurrences in the classroom send signals to students, and those signals have impact—regardless of intent.

Sociomathematical Norms

Social norms are community specific—not subject specific. Sociomathematical norms are specific to a *mathematics* learning community and regulate the community's communication about and participation with the subject of *mathematics*. So, while a social norm about respectful argumentative discussion might define how community members (students) debate/critique the legitimacy of a solution to a problem, only a sociomathematical norm can define what constitutes mathematical legitimacy in that community, what defines a contradiction (such as a counterexample), or otherwise (Cobb & Yackel, 1996; Rasmussen et al., 2003; Yackel & Rasmussen, 2002). Notice that the social norm that dictates debate structure can be transferred to other communities studying other disciplines, whereas the sociomathematical norms that dictate mathematically sound claims only apply within the context of a mathematics learning community. Aside from being subject specific, sociomathematical norms are much like social norms in that they are negotiated and renegotiated by the community in order to organize and regulate the activities that lead to learning (Cobb & Yackel, 1996; Rasmussen et al., 2003). The difference, again, is that sociomathematical norms regulate the activities at the mathematical level whereas social norms regulate the activities at the surface level.

Let's consider this example: *Attend to Precision* (for younger students, the norm might be work carefully—use the right words, symbols, and units of measurements). The intent of this sociomathematical norm is grounded in the language and notation used to describe mathematical ideas and claims. These are social adherences to convention,

Sociomathematical norms: norms that are specific to a *mathematics* learning community and regulate the community's communication about and participation with the subject of *mathematics*.

not purely mathematical. And while this likely falls under the umbrella of explaining one's thinking, it deserves its own focus in that it normalizes what it means to be mathematically unambiguous (and conversely, it also sheds light on what is mathematically ambiguous). A mathematical claim can only be considered if it can be understood. Therefore, it is critical that students develop the content language and mathematical sophistication to clearly communicate their thinking in a *precise* manner. One might describe this act as *leaving the reader/listener no choice but to interpret our claim the way we intend.* In other words, if the student attends to precision, there is no room for interpretation. To build a need for this development in students, teachers might consider communicating a message to their students when reading their work along the lines of, "*I know exactly what you mean (and I agree with you!). But what if I wasn't the reader? What if I wasn't the teacher who knew you well enough to read between the lines and infer what *you* really mean? How could you ensure that *you are heard* regardless of the audience?*" The implied social norm in this response, that the teacher's role is that of their advocate, sends the signal that the students *are* mathematicians and that their ideas *are* worth being made public. This shapes their agency. It follows that these ideas should be interpreted as intended by the authors—*them*. Hence, the social norm is used in service of the sociomathematical norm.

> Attending to precision leaves the reader or listener no choice but to interpret our claim the way we intend.

Classroom Mathematical Practice

A classroom mathematical practice is something that begins as any other claim, answer, or solution method and eventually gets adopted as common knowledge or common *practice* by the community (Rasmussen et al., 2003). A number of these types of classroom practices fall under the umbrella of Common Core Standard for Mathematical Practice 8, *Look for and express regularity in repeated reasoning* (National Governors Association Center for Best Practice, Council of Chief State School Officers, 2010). This standard for mathematical practice is all about finding generalizations and often results in reusable formulas or patterns. One such pattern that (based on the classroom structure) might begin as a discovery and student claim and later become common practice is the difference of two squares factorization pattern. With a well-designed task, students could discover that $(x + y)(x - y) = x^2 - y^2$ and that the factors of $x^2 - y^2$ are always $(x + y)(x - y)$. While this discovery might begin as novel, it will almost certainly become a regular practice of the classroom community, as its usefulness is apparent in its repeated use. In elementary classrooms, students might recognize through repeated practice and attempts that multiplying a number by 10 shifts its place-value over one to the left and puts a 0 in the ones place. When this fact becomes expected knowledge and no longer requires justification by the community, it has become a classroom mathematical practice.

As teachers, we have to be cautious with what becomes taken as given and how that process plays out in our classrooms. Are concepts becoming taken as given because of their frequent use and the community's shared understanding, or because a dominant voice (possibly ours) insists on their validity and thus brushing over? If the latter becomes our process of developing classroom mathematical practices, we run the risk of miscommunicating mathematics as a faith-based endeavor. On the flip side, we must also allow these practices to develop if students are to truly assume an ownership role in mathematics. If, as teachers, we continue to demand that students justify and validate concepts on every mentioning well past the point of generalization, we run the risk of miscommunicating mathematics as something that cannot be personalized and simplified. For is it really *ownership* if you must always rebuild it from scratch to prove your ownership?

Explicitly Developing Social Norms With Social Learning Intentions

The impact of social norms should not be understated—the ways in which we interact with each other serve to define our beliefs about mathematics itself as well as our own place within it. In other words, social norms serve to situate us within mathematics. So then, how should we as teachers work to shape the norms of interaction within our classroom? Further, how *can* we even go about doing such a task; what vehicles exist to intentionally develop social norms? One such tool, social learning intentions, can be designed and leveraged to do just this. In order to discuss social learning intentions, however, let's back up a step and frame them within the larger construct of learning intentions generally.

Learning intentions are stated goals of *learning* (not doing) around which we design our learning experiences and tasks. Learning itself is complex, and we recognize it to have pure content components, as well as language demands nested in social interactions. This is why, in practice, learning intentions should be broken into three components: content, language, and social.

- **Content learning intentions** describe what learning can be accomplished toward grade-level standards in a given day. In mathematics, content learning intentions answer the question for students, *What is the mathematics I am supposed to **use** and **learn** today?*

- **Language learning intentions** define the language demands of the day's lesson and address how students will demonstrate their thinking with language. Language learning intentions answer the question for students, *How should I communicate my thinking today?*

- **Social learning intentions** speak to how students will work with one another and also provide a space to address the social-emotional demands of the day. Social learning intentions answer the question for students, *How should I interact with my learning community today?* Further, social learning intentions are a tool to mold classroom culture, community, and social norms early in the year and leverage classroom culture, community, and social norms later in the year.

> Social learning intentions are a tool to mold classroom culture, community, and social norms early in the year and leverage classroom culture, community, and social norms later in the year.

Writing social learning intentions and planning them into our lessons begins with identifying what from the affective domain we want to target for development in our students in a given day. What will our students need to *do* socially or emotionally in order to meet the demands of the day? Further, are these skills that they arrive with, or do they need to be deliberately modeled and practiced? In other words, are we developing or leveraging social norms? Clearly, this all begins with knowing our students and classroom community.

CHAPTER 2 SUCCESS CRITERIA CALLOUT:

☐ I can use social learning intentions as a vehicle to develop and leverage social norms in my classroom.

Let's imagine an example from a fourth-grade class. In order to promote a social norm around valuing diversity of ideas and learning to work with many different partners, the teacher must create opportunities for this to occur. So then, let's assume that the teacher implements strategic heterogeneous grouping strategies at the start of each unit of study to create these opportunities. This action sets the stage for the social norm to develop but absolutely does not guarantee the direction in which such development will occur. In other words, the students are most definitely about to develop norms around how they navigate diverse ideas and work with partners whom they don't choose, but if the teacher intends these norms to be in favor of mathematical ownership, they better keep their hand on the scale. The teacher might consider implementing a social learning intention on the first day of this new unit saying something along the lines of, *We are getting used to our new groups by asking for and offering help and support when needed.* By elaborating this social goal for the day, the teacher is acknowledging the emotional reality that getting used to working with new people is necessary. Further, the teacher is providing an avenue through which students might engage in such acclimation and learn to value one another—by asking for and offering help and support. This emphasis on peer-to-peer assistance serves to reinforce additional social norms about the distribution of mathematical authority (it's not just the teacher's!) and helpful classroom interactions.

Let's be clear—simply posting or stating a social learning intention at the onset of a lesson does not immediately manifest a shared social norm in a classroom community (again, imagine if it were that easy?). Much like content and language learning intentions, social learning intentions are stated goals to be elaborated on and revisited throughout a lesson. Teachers should discuss social learning intentions at the beginning of a lesson to ensure students understand their meaning; they should consider modeling their meaning through exemplars/work samples, a think-aloud, or an acted example; and they should reference them throughout the day as a moderating force for assessing interactions. Further, tasks should be designed or chosen that support learning intentions by providing students the opportunities to work toward those stated goals.

Since classroom culture is dynamic, social norms are shifting in nature. When writing social learning intentions to shape social norms, it can be helpful to assess the current state of social norms in your classroom and identify where you would like to shift them. What, for example, are the existing social norms surrounding mathematical autonomy and self-efficacy in your classroom, and how do they compare to where you would like them to be? Consider the *Assessing Social Norms Teacher Questionnaire* in Figure 2.1 as a place to begin understanding the current state of social norms in your classroom. In addition to helping manifest existing social norms, this tool provides suggested social learning intentions to shift social norms toward a more productive space in each domain. Though extensive, this tool is by no means exhaustive in identifying social norms, domains of social norms, or social learning intentions. Rather, this tool is intended to expose the reader to different examples of social learning intentions, their uses, and their varying structures as an invitation to join the practice. I hope that these exemplars serve as a grounding point for you to write your own social learning intentions that take into account your specific context and meet the needs of your students.

Figure 2.1   Assessing Social Norms Teacher Questionnaire

ASSESSING SOCIAL NORMS TEACHER QUESTIONNAIRE: ELEMENTARY

Select a value that best represents where the related social norm in your classroom currently lands on each spectrum. Social norms are considered consensus entities, so choose values based on where you believe a *majority* of your students would agree.

DOMAIN 1: MATHEMATICAL AUTONOMY AND SELF-EFFICACY

We have to ask the teacher to determine what is mathematically valid.	−4 −3 −2 −1 0 1 2 3 4	We can discover math, so we should try to figure out what is mathematically valid before asking the teacher.

Social Learning Intentions to shift or reinforce this norm:

- To remind each other of the math power we have together. *We know this stuff!*
- To be curious about new problems. *Let's try to figure it out!*
- To actively participate in all activities.
- To use what we know to make sense of new ideas or problems.
- To be comfortable with new things. Give it a shot—what's the worst that could happen?

When we don't understand something, we just wait for the teacher to explain it.	−4 −3 −2 −1 0 1 2 3 4	When we don't understand something, we can discuss it and make sense of it as a community.

Social Learning Intentions to shift or reinforce this norm:

- To look for and talk about big ideas and connections between this lesson and previous lessons.
- To learn from failure when things don't make sense . . . yet.
- To practice until we feel comfortable and problems feel familiar.

DOMAIN 2: SELF-REGULATION AND MOTIVATION

We learn as much as our teacher decides to teach us.	−4 −3 −2 −1 0 1 2 3 4	We are in control of how much we learn.

Social Learning Intentions to shift or reinforce this norm:

- To be prepared for learning and ready to ask questions.
- To recognize which things you will need to spend extra time to learn and be willing to practice more.
- To reflect on what *worked* and what *did not work* with previous learning. *What should you keep and what should you change?*

DOMAIN 2: SELF-REGULATION AND MOTIVATION (continued)		
We give up when math gets hard.	−4 −3 −2 −1 0 1 2 3 4	We accept challenges and recognize that hard work is required for learning.

Social Learning Intentions to shift or reinforce this norm:

- To practice the focus needed to learn new things and practice new problems.
- To recognize that you have to work hard when learning doesn't come easily. But we are here to help!
- To build math stamina.
- To be patient and ready to practice in order to master new skills.
- To practice positive habits of mind. This might be hard, but we can get there.
- To practice positive self-talk when trying new problems. Confidence makes a difference!

DOMAIN 3: SENSE OF URGENCY		
We don't really care if we get through a whole lesson in a day.	−4 −3 −2 −1 0 1 2 3 4	We want to learn everything we are supposed to in this class.

Social Learning Intentions to shift or reinforce this norm:

- To transition smoothly between individual, group, and whole-class activities.
- To use class time wisely and with purpose.
- To work hard and finish strong!
- To commit to success each day and all year.

DOMAIN 4: COMMUNICATING AND ORGANIZING MATHEMATICS		
We know math when it makes sense watching the teacher do it.	−4 −3 −2 −1 0 1 2 3 4	We know math when we can decipher and communicate it with precision.

Social Learning Intentions to shift or reinforce this norm:

- To make sense of each piece of the problem.
- To stay organized in our thinking and our work.
- To remember that it's okay to work slowly when we need to so that we don't make silly mistakes.
- To use math language to share our thoughts and work precisely.

(Continued)

(Continued)

DOMAIN 5: WORKING WITH OTHERS		
We don't work together, we do math alone.	−4 −3 −2 −1 0 1 2 3 4	We value different ideas and work with many different partners.

Social Learning Intentions to shift or reinforce this norm:

- We work in groups and those groups can change. It's important to ask questions and offer help when needed.
- We work together to discover new ideas and try new things.
- We share our thinking with others and stay open to their ideas.
- *Dare of the day*: Go check in with one person who you don't normally work with about one of the classwork or homework problems.
- We practice explaining challenging problems and ideas to someone else.

We only need to worry about our own learning.	−4 −3 −2 −1 0 1 2 3 4	We are here to support and build each other up, not tear each other down.

Social Learning Intentions to shift or reinforce this norm:

- To help our classmates learn complex ideas.
- To listen and be respectful when classmates share their thinking.
- To encourage everyone in class to succeed at the highest level possible.
- To support one another and use kind and encouraging words, even when learning is hard.

online resources
Available for download at **https://companion.corwin.com/courses/whosemathisit**

ASSESSING SOCIAL NORMS TEACHER QUESTIONNAIRE: SECONDARY

Select a value that best represents where the related social norm in your classroom currently lands on each spectrum. Social norms are considered consensus entities, so choose values based on where you believe a _majority_ of your students would agree.

DOMAIN 1: MATHEMATICAL AUTONOMY AND SELF-EFFICACY

We have to ask the teacher to determine what is mathematically valid.	−4 −3 −2 −1 0 1 2 3 4	We can discover math, so we should try to figure out what is mathematically valid before asking the teacher.

Social Learning Intentions to shift or reinforce this norm:

- To remind one another of our collective mathematical authority—_we know this stuff!_
- To approach each task inquisitively—_try to figure it out!_
- To actively participate in knowledge–generating, discovery activities.
- To actively pursue and leverage overlap in old and new content. _Use what you know to make sense of the new._
- To develop a comfort with the unfamiliar . . . give it a shot—what's the worst that could happen?

When we don't understand something, we just wait for the teacher to explain it.	−4 −3 −2 −1 0 1 2 3 4	When we don't understand something, we can discuss it and make sense of it as a community.

Social Learning Intentions to shift or reinforce this norm:

- To pursue generalizations and connections between this and previous lessons through open discussion.
- To practice the algebraic skill of _chunking_ as a tool to navigate complex problems. (Zoom out for a sense of direction; zoom in to progress.)
- To harness the power of disequilibrium: _this doesn't make sense . . . yet_. What makes mathematicians different from others is that we've gotten really comfortable with seeking and learning from failure. We are _triggered_ by the unknown in that we seek to _know_ it.
- To seek the comfort of familiarity, and when you can't find it, create it through repetition—_practice_.

DOMAIN 2: SELF-REGULATION AND MOTIVATION

We learn as much as our teacher decides to teach us.	−4 −3 −2 −1 0 1 2 3 4	We are in control of how much we learn.

Social Learning Intentions to shift or reinforce this norm:

- To arrive to lab with classwork done, homework started, and questions ready.
- To reflect on our preparedness for the last exam as we start a new unit.

(Continued)

(Continued)

DOMAIN 2: SELF-REGULATION AND MOTIVATION (continued)

- To recognize where our learning has fallen short and take action to bridge the gap. Advocate for your learning; be ready to work for what didn't come naturally.

- To recognize which areas of the content you will need to spend extra time to learn, and to advocate for your learning inside and outside the classroom.

- To reflect on which preparation methods *worked* and which perhaps *did not* on the last exam. *What should you keep, and what should you change?*

We give up when math gets hard.	−4 −3 −2 −1 0 1 2 3 4	We accept challenges and recognize that hard work is required for learning.

Social Learning Intentions to shift or reinforce this norm:

- To practice the focus required for understanding and navigating complex problems.

- To recognize that new learning requires heavy lifting, and that the heavy lifting can't be done by anyone but you (but we are here to spot you).

- To practice the stamina required for advanced math—it's a marathon!

- *Use, use your powers; what now costs you effort will in the end become mechanical.— Georg C. Lichtenberg*

- To demonstrate the patience and practice required in order to master new skills.

- To practice the intense work ethic required for the highest levels of success. If you want to *reach* the top, reach *for* the top.

- *We are what we repeatedly do. Excellence, then, is not an act, but a habit.—Aristotle*

- To practice the positive habits of mind that have gotten you this far. This content might feel a little foreign and look a little new, but we've won those battles before!

- To practice positive self-talk when approaching complex problems—*including word problems.* Confidence makes a difference!

DOMAIN 3: SENSE OF URGENCY

We don't really care if we get through a whole lesson in a day.	−4 −3 −2 −1 0 1 2 3 4	We want to learn everything we are supposed to in this class.

Social Learning Intentions to shift or reinforce this norm:

- To transition smoothly between individual, group, and whole-class activities.

- To effectively use each moment of class time—we are running out!

- To respond to the sight of the finish line with an all-out academic sprint—finish strong!

- To dedicate ourselves to success during the final stretch of this course.

DOMAIN 4: COMMUNICATING AND ORGANIZING MATHEMATICS		
We know math when it makes sense watching the teacher do it.	−4 −3 −2 −1 0 1 2 3 4	We know math when we can decipher and communicate it with precision.

Social Learning Intentions to shift or reinforce this norm:

- To reconcile what we have understood conversationally with the discipline required for pure mathematical precision.

- To make sense of each piece of complicated abstract notation.

- To stay organized in your thinking so that you may keep track of the goals of complex problems. *Am I differentiating, or antidifferentiating? Which direction am I going?*

- To make sense of each piece of complicated abstract notation.

- I am seeking to understand complex notation, rather than letting it intimidate me.

DOMAIN 5: WORKING WITH OTHERS		
We don't work together, we do math alone.	−4 −3 −2 −1 0 1 2 3 4	We value different ideas and work with many different partners.

Social Learning Intentions to shift or reinforce this norm:

- We are getting acclimated to our new groups by asking for and offering assistance when needed.

- We are working toward a common discovery.

- To share our thinking with others while also being open to their ideas.

- *Dare of the day*: Go check in with one person you don't normally work with about one of the classwork or homework problems.

- To practice explaining complex concepts and ideas to someone else. *The standard of knowing is explaining your thinking to someone new.—Dr. Frey*

We only need to worry about our own learning.	−4 −3 −2 −1 0 1 2 3 4	We are here to support and build each other up, not tear each other down.

Social Learning Intentions to shift or reinforce this norm:

- To help our peers navigate complex ideas.

- To collaboratively and respectfully challenge the thinking of our peers—and to share nicely. :)

- To hold one another accountable, using growth-producing language, to our attention to precision—grammar police style.

- To support one another in meeting the high expectations of this class.

 Available for download at **https://companion.corwin.com/courses/whosemathisit**

Sociomathematical Norms: For Better or Worse

Sociomathematical norms develop in a community whether we intend them to or not. The way that we model engaging with mathematics, the way we communicate about mathematics, what we accept as a complete response, and a multitude more of our actions as teachers act to establish or reinforce the nature of mathematics in our classrooms. This section argues that our teacher clarity and subsequent intentionality should extend to all aspects of instruction and community interaction. To illustrate this, let's look at an eleventh-grade classroom community and consider how a sociomathematical norm might be shaped in the following situation.

CHAPTER 2 SUCCESS CRITERIA CALLOUT:

☐ I can take a proactive approach to shaping productive sociomathematical norms in my classroom.

Imagine the teacher is scanning the classroom as students work on the following task toward the close of a class period.

The height of a rock thrown into the air is given by $h(t) = 32t - 16t^2$ feet, where t is measured in seconds.

(a) Calculate $h(1)$ and give a practical interpretation of your answer.

(b) Calculate the zeros of $h(t)$ and explain their meaning in the context of this problem.

As the teacher walks the classroom, he notices a lot of emphasis on the calculation portions of this task while students largely seem to be ignoring the practical explanations of their results. Seeing this, he feels compelled to prompt his students and announces generally as students continue to work, "Let's remember to read what the entire problem is asking. . . ." Continuing his scan while intermittently offering extra support to certain groups, he still notices students recording lots of numbers without a context. He decides to escalate his scaffolding from a prompt to a cue and tells his students, "Don't forget to include units in your answers!" This seems to give pause to many students and even causes some audible grunts as they revisit their *believed-to-be* completed work.

As the day comes to a close, the teacher collects the students' work and notices that his cue seems to have *worked*, in that only a few students turned in contextless calculations. However, while most students in fact included units in their answers, some mixed up seconds and feet. Further, those that did correctly label their results with units did so as their only ode to the original context. In other words, their use of units was their only evidence of a practical interpretation of their calculations. This pivotal data provides a decision-point for the teacher that holds implications beyond the students' adherence to the instructions of this specific task. Students in this community clearly share a sociomathematical norm that *mathematical completeness emerges as calculations conclude.* The problem, I would argue, with a norm like this is multifaceted and begs us to revisit the Strands of Mathematical Proficiency.

PROS: Raw calculations, such as those completed by students initially, can help us develop procedural fluency—something of the utmost importance in mathematics if we are to develop the efficiency required to advance in the subject.

CONS: By not thoroughly revisiting the context, or by doing so using incorrect units, students provide evidence that their conceptual understanding on the topic might be lacking. Note that the task specifically *told* students what procedures to engage in. By simply engaging in those procedures in the symbolic space and not thoroughly connecting the results to the contextual space, the students did not demonstrate whether they were recognizing mathematical equivalence in different representations (from symbolic → words, in this case). Further, by virtue of telling students what to calculate and then asking for contextual interpretation, a task like this acts as a scaffold toward developing strategic competence. Strategically competent students will engage in this process of mathematization independently, which is something that seems less likely for this current group if they appeared to ignore or at least undervalue the call for contextual interpretation. Again—this data marks a clear decision-point for this teacher as he works to tilt the scales toward sociomathematical norms that promote success and ownership in mathematics.

The teacher wants this community to redefine mathematical completeness with a greater emphasis on mathematical sophistication. Namely, he wants his students to value revisiting the context in problems as a means of making thorough conceptual connections between mathematical representations. This emphasis on deep understanding can serve as a springboard toward transferring mathematics to novel situations—the capstone of mathematical ownership. However, as I mentioned at the onset of this chapter, culture (of which norms serve to define), cannot simply be *set* by the teacher. Norms must be negotiated and agreed upon by the members of a community. Thus, students must participate in developing and refining sociomathematical norms—they have to believe they are *choosing* them if they are to impact their sense of autonomy and mathematical ownership. Therefore, rather than telling the students what constitutes a mathematically complete response, the teacher wants to facilitate a conversation around the following questions: *What should be considered mathematically complete? What are our options? Why should we choose a certain one?* This approach allows ownership to develop around the norm itself, which feeds into beliefs about mathematics and one's own agency.

With the aforementioned goal in mind, the teacher designed a follow-up task for the next day by collecting some sample student responses for comparison and discussion. He chose two typical responses—both with correct calculations but only one that reported units—and rewrote them in his own handwriting. Then—and this is where the scale-tilting comes in—he fabricated a third sample that included a more thorough and sophisticated response. All three comparative samples are shown in Figure 2.2.

Figure 2.2 Comparative Sample of Two Student Responses and a Leading Third

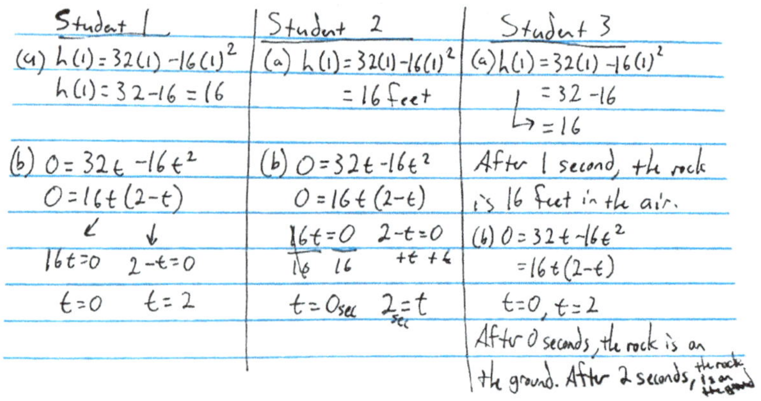

Student 1	Student 2	Student 3
(a) $h(1) = 32(1) - 16(1)^2$	(a) $h(1) = 32(1) - 16(1)^2$	(a) $h(1) = 32(1) - 16(1)^2$
$h(1) = 32 - 16 = 16$	$= 16$ feet	$= 32 - 16$
		$\rightarrow = 16$
(b) $0 = 32t - 16t^2$	(b) $0 = 32t - 16t^2$	After 1 second, the rock
$0 = 16t(2 - t)$	$0 = 16t(2 - t)$	is 16 feet in the air.
↙ ↓	$16t = 0$ $2 - t = 0$	(b) $0 = 32t - 16t^2$
$16t = 0$ $2 - t = 0$	$\frac{}{16}$ $\frac{}{16}$ $+t$ $+t$	$= 16t(2 - t)$
$t = 0$ $t = 2$	$t = 0$ sec $2 = t$ sec	$t = 0, t = 2$
		After 0 seconds, the rock is on
		the ground. After 2 seconds, the rock is on the ground

The accompanying plan for this artifact is centered on comparison and analysis. Essentially, the teacher intends to calibrate his students as task-scorers through facilitated discussion. In groups, the teacher plans to ask students a series of questions around which consensus—and ultimately a sociomathematical norm—could be built. His ordered questions, desired student responses, and planned follow-up questions to elicit those responses for each are as follows:

Question 1: What is the same or similar between all three responses?

Desired Responses and Accompanying Follow-Up Questions:

- Same calculations; Same methods; *Did anyone do this problem differently?*

- Same Answers; All correct; *Who got the calculation portion of the problem correct?*

Question 2: What is different between all three responses?

Desired Responses and Accompanying Follow-Up Questions:

- One response has no units, the other two have units; *How does the first response compare to the other two? What makes it distinct?*

- One response uses complete sentences to rewrite their answers; *What is unique about the third response?*

- One response puts the answer back into the context of the problem; *Which responses remind us what the original problem was about?*

Question 3: What can we fairly determine each student *knows* based on their written response?

Desired Responses and Accompanying Follow-Up Questions:

- All students know how to calculate the answers; *Is it fair to say all students showed an understanding of how to make these calculations?*

- There is no evidence that Student 1 understands how the calculations relate to the context—they might, but we don't know that; *Can we fairly say that Student 1 understands how their calculations relate to the context? Why, or why not?*

- Student 2 relates their answer back to the context with correct units, but that might not be strong evidence for their understanding of those connections; *How well can we prove that Student 2 understands how their calculations relate to the context?*

- Student 3 thoroughly relates their answer back to the context and leaves no doubt that they understand the connections between the context and their calculations; *If somebody said that Student 3 did not understand how their calculations related to the context, how would you rebut them?*

Question 4 (Reflection): What should we consider mathematically complete? Why?

Desired Responses and Accompanying Follow-Up Questions:

- We should show all of our work so it is clear that we understand how to make calculations; *How can we prove that we understood all of our calculations?*

- We should write our answers in sentences that go back to what the problem is talking about to show that we understand how our calculations relate to the real world; *How can we prove that we understand how our calculations relate to the problem's context?*

This illustrative example is intended to highlight some key features of sociomathematical norms.

1. **Sociomathematical norms exist, develop, and persist with or without our input and shaping.** If the teacher chose not to act at this decision-point, he could be reinforcing the existing norm that mathematical completeness is congruent with completed calculations.

2. **Teachers can absolutely influence the direction that sociomathematical norms develop.** This begins with noticing where they exist in their current state and designing opportunities to renegotiate their contours.

3. **Sociomathematical norms are communal and teachers are but one member of the learning communities** over which we preside— students are the central (and majority) players in our classrooms.

With these features in mind, we must make students active participants in negotiating and renegotiating sociomathematical norms. If they are to develop mathematical ownership then they must be provided opportunities to appreciate and value the constructs of sophistication, completeness, and efficiency that comes with mathematical understanding. We cannot simply urge our students to comply into ownership—as no such route exists. For ownership, at its core, is about independence.

Communicating and Modeling Choice in Mathematics

What characteristic better illustrates ownership than the freedom to make choices? Home ownership allows the owner to paint their interior walls whatever color they choose. Restaurant owners can choose what to offer on their menus, their décor, and how to set their pricing. Ownership also tends to come with responsibility, however, as choices have consequences. The homeowner who decides to paint his granddaughter's bedroom hot pink with unicorn wallpaper trim, for example, might turn off some prospective buyers when it comes time to sell (true story!). The menu decisions made by restaurant owners can act to exclude or include potential clientele-bases, their décor choices affect how they brand their business (and thus who they reach), and their pricing can determine the likelihood of developing repeat business—or, attracting regulars. Thus, the freedom to make decisions is both the trademark and burden of ownership. Decision-making, for better or worse, is also the defining characteristic of mathematical ownership.

Let's consider some key features of mathematical ownership. First, we can't make choices if we don't have options to choose from. So, mathematical ownership encompasses expanding our problem-solving repertoires to include multiple options. Second, as highlighted by the homeowner and restaurant owner examples, decision-making has direct consequences. In mathematics, these consequences often manifest as differences in process. So, what is our guiding light, as mathematicians, that can help us anticipate these consequences and make decisions with intention?

There are few concepts more consequential in secondary mathematics than the slope of a line. This concept, analogous to the rate of change between bivariate pairs, is a through line (no pun intended) from pre-algebra to calculus. What typically begins as using counting methods to note the change in two quantities over certain intervals evolves into calculating and comparing differences using a ratio. We generalize and formulize this

CHAPTER 2 SUCCESS CRITERIA CALLOUT:

☐ I can promote choice in mathematics to build student ownership.

concept in algebra in order to develop the fluency required to explore it in depth in calculus—where we take it to the extreme by developing the limit definition of the derivative, which shrinks the difference between subsequent x-values to zero. Table 2.1 illustrates this evolution of slope over a student's secondary and postsecondary journey.

Table 2.1   Secondary and Post-secondary Evolution of Slope

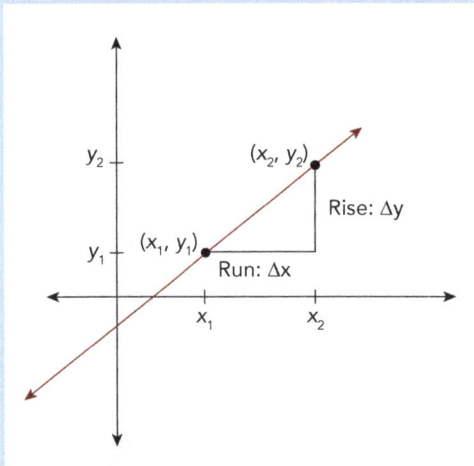

Slope of a Line:

$$m = \frac{rise}{run} = \frac{change\ in\ y}{change\ in\ x}$$

$$= \frac{\Delta y}{\Delta x} = \frac{y_2 - y_1}{x_2 - x_1}$$

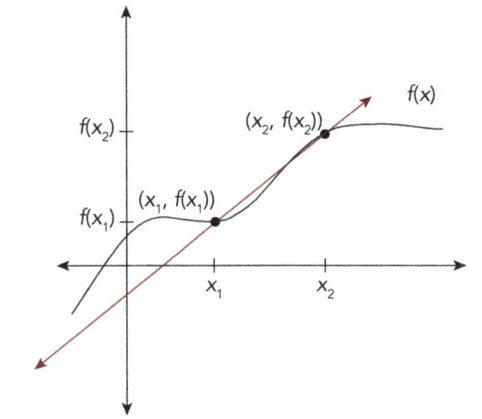

Average Rate of Change :

$$Slope\ of\ the\ Secant\ Line = m_{sec} = \frac{f(x_2) - f(x_1)}{x_2 - x_1}$$

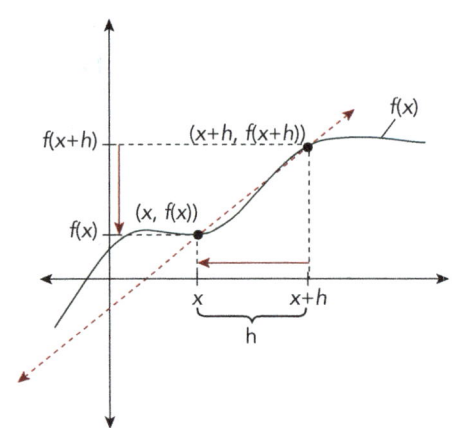

Derivative of $f(x)$:

$$f'(x) = \frac{dy}{dx} = \lim_{h \to 0} \frac{f(x+h) - f(x)}{h}$$

Take a pause from reading, and complete the following abstract procedural exercise on calculating the slope—one of a multitude of similar exercises you and your students have (or will) likely encountered in your mathematical journeys. (For some of you, you may not have solved a problem like this in a long time. I encourage you to give it a try as best you can remember.) Even though this is likely securely-held content, I encourage you to write every step of your process including setup, calculations, and conclusion.

Calculate the slope between the points (2, 3) and (5, 7) using the slope formula:

$$m = \frac{y_2 - y_1}{x_2 - x_1}$$

How'd it go? Did you get $\frac{4}{3}$? Great if you did, but that's not the interesting part. How did you set the problem up? I would wager to say that most readers set the exercise up as follows:

$$\frac{y_2 - y_1}{x_2 - x_1} = \frac{7 - 3}{5 - 2} = \frac{4}{3}$$

If this is the case for you, why did you set it up this way? Were you following the perceived order of x_1, y_1, x_2, y_2 as presented in the problem? Was your choice totally random? Or, was your choice based on foreseen consequences? Here's the bottom line—in calculations like this (and in many others in mathematics), choices of a starting point exist in that there are multiple valid paths to a mathematically correct response. Sometimes this is arbitrary and provides autonomy to the user, in other words, the user can choose how to start with little to no consequence. In other instances, however, the direction in which we exercise our mathematical autonomy determines the difficulty (and even *possibility*) of the path ahead. What if I had presented the pairs of points in the prior exercise in the reverse order (i.e., Calculate the slope between the points (5, 7) and (2, 3) using the slope formula)? How would you have set up your calculations in that case? Many would argue to still lead with setting $(x_2, y_2) = (5, 7)$ and $(x_1, y_1) = (2, 3)$, because the subtraction of 7 − 3 is generally perceived as simpler than 3 − 7, and likewise with 5 − 2 versus 2 − 5. The bottom line is that users are empowered to make choices when they possess a sense of direction and can do so with intention. What would your students do? How would they set this exercise up? Try it—give them this exercise as an exit ticket, and see what they choose.

Procedural fluency is not just about recitation—it is about authority and ownership of decision-making.

We need to be careful, as teachers, not to let the mathematician-within-us consider this as trivial. We need to talk about these choices with our students and not just brush them out of the way as if all that matters is the inconsequential slope of the line between these two points. Procedural fluency is not just about recitation—it is about authority and ownership of decision-making. Consider the following task to further illustrate this concept within the context of calculating slope. Based on student responses to the exit ticket suggested prior, you might consider facilitating a whole-class conversation after students develop responses to this task in small groups.

Three students are debating about how to calculate the slope of a line between two points. Lizbeth and Naima agree on the slope itself but argue about the process. Matthew disagrees on the slope *and* the process. Who is right?

Problem: Calculate the slope between the points (–2, 4) and (5,–6).

Lizbeth's Work:

$$\frac{y_2 - y_1}{x_2 - x_1} = \frac{-6-4}{5-(-2)} = \frac{-10}{7} = -\frac{10}{7}$$

Naima's Work:

$$\frac{y_2 - y_1}{x_2 - x_1} = \frac{4-(-6)}{-2-5} = \frac{10}{-7} = -\frac{10}{7}$$

Matthew's Work:

$$\frac{y_2 - y_1}{x_2 - x_1} = \frac{-6-4}{-2-5} = \frac{-10}{-7} = \frac{10}{7}$$

When facilitating discussions around choice in mathematics, we should be careful not to reinforce the norm that "all that matters is that you get the right answer." This norm, I would argue, overemphasizes product over process and does not wholly promote success in mathematics. Rather, we should emphasize procedural decision-making by asking "whose math is it?" to highlight each decision-point. Additionally, tasks such as these provide opportunities to develop sociomathematical norms around preference and student-perceived ease. To be blunt about it, I use tasks like these in my classroom to promote the sociomathematical norm that *math is about being LAZY*. This norm is meant to help students regulate their procedural decision-making by thinking a few steps ahead and choosing what they consider to be the path of least resistance. This, I would argue, promotes ownership with intention.

We should also be direct that this task isn't meant to be an extraordinary or novel example—quite the contrary. This example is meant to highlight that developing mathematical ownership, especially at the whole-class level, is greatly impacted by our own routine daily disposition toward mathematics as teachers. Talking about our thinking, modeling our decision-making, and highlighting where students have the autonomy of choice serves to empower students to take control of the mathematics. This is especially impactful when we find ways of addressing this even in routine tasks, such as the previously mentioned slope example. The more opportunities we find to ask students *Whose math is it?* to highlight a mathematical choice, and to promote that math is about being efficient (or *lazy*), the more credibility we gain as messengers of a personal subject. Thus, as stated at the onset of this section, the task of modeling choice in mathematics is about both enhancing our students' understanding of the subject as much as it is about messaging *what* and *how* mathematics *is*. For mathematics exists not only as a beautiful collection of independent and objective truths but also as a personalizable tool.

> Talking about our thinking, modeling our decision-making, and highlighting where students have the autonomy of choice serves to empower students to take control of the mathematics.

Building Agency With Discursive Positioning Moves

Imagine you are participating in a professional development training and you perceive that you have relatively low familiarity with the topic at hand compared to your colleagues—how likely are you to openly share your thinking with the whole group? Now, imagine the whole event is happening in a language other than your first. How might this affect your participation? Further, what messages might you be receiving or self-generating about your status as a contributor to your learning community? Are you more or less likely to consider your early ideations as valid?

Learning requires us to embrace vulnerability. We must expose our current conceptions of ideas, reconcile previously held understandings (or lacks thereof) with new information, and possess sufficient self-concept to believe we have a shot at figuring things out. In other words, we have to find comfortable footing in discomfort and believe that *trying* is not futile. And, as illustrated in this section's opening scenario and questions, there are often additional layers of complexity that make this all the more challenging. School isn't easy, and we aren't necessarily saying that it should be, but it should most definitely be accessible. Many students in our classrooms arrive with varying degrees of unfinished learning from previous years. Many students arrive in the process of learning the primary language in which school is taught. Many students arrive with negative dispositions toward math, themselves within math, or both. Many students don't see this as *their math*. We, as teachers, must work to ensure that these challenges don't preclude students from participating and accessing mathematical learning and success. We've discussed many tools to help promote mathematical ownership in this chapter. The tool highlighted in this final section, discursive positioning moves, serves to address the challenges (and others like it) highlighted here by situating students as agentive problem solvers (Turner et al., 2013).

Discursive positioning is the verbal process through which students are regulated in a community—from teacher-to-student and from student-to-student—and assigned various statuses. The way community members respond to our contributions define the quality of those contributions, our position as a contributor, and our status and role within the community. And, as with many of the constructs covered in this chapter, discursive positioning happens with or without the influence or intention of teachers. Thus, the clarity afforded by the definition of mathematical success should inform our intentional use and regulation of discursive positioning in our classrooms. In other words, let's position our students as *owners* of mathematics. To further emphasize this, studies have shown that the nature of discursive positioning becomes part of a learning community's culture—as clear norms around how students position one another during discussions develop over time (Yoon, 2007, 2008). This is especially evident and impactful with how English learners are situated as valued and valid contributors to discussions. So, it behooves us to understand the types of discursive positioning moves so that we can simultaneously leverage them in our classrooms and model their usage for our students.

> ### 🔍 CHAPTER 2 SUCCESS CRITERIA CALLOUT:
>
> ☐ I can use discursive positioning moves to situate all students as productive contributors during whole-class discussions.

Discursive positioning: the verbal process through which students are regulated in a community—from teacher-to-student and from student-to-student—and assigned various statuses.

In mathematics classrooms there are three categories of discursive positioning moves (Turner et al., 2013):

(1) Explicit statements that validate students' reasoning

(2) Invitations to share, justify, or clarify thinking that position students as competent problem solvers

(3) Invitations for peers to respond to a student's idea in ways that position the idea as important or mathematically justified

Explicit statements that validate student's reasoning, involves two typical methods: restating and revoicing. Restating works like an amplifier. Teachers can literally repeat or paraphrase a student's words in order to add emphasis as an act of focusing attention to a single contribution/contributor (e.g., "Ok, wow, let's take a moment and think about this. Lina just said, 'I looked at the digit in the 10s place to help me figure out that 73 is larger than 43; 7 tens is more than 4 tens.' That's big.") These types of statements give pause to a discussion and force students to take time to process the contributions of their peers. This peer-processing is impossible to do without recognizing one's peers as contributors, thus the *positioning* of discursive positioning manifests. Revoicing works very similarly on the meta-level of facilitating discussions, though it has differences in operation. Revoicing is the act of editing-then-amplifying a student's response by perhaps adding some connective logic, formal terminology, or anything else to help communicate the student's reasoning—all the while assigning ownership to the student (e.g., "Ok, Diego is making a very important point here. Diego, correct me if I am not saying this right, but you are basically arguing that since the first function is constantly multiplying, while the second function is just constantly adding, that the first function will grow faster and get to 100 first, right? Are we all understanding Diego's point?"). This intentional rephrasing allows the teacher to elevate contributions that might be developed in concept, though not yet completely in communication. Revoicing effectively expands the potential pool of contributions (and thus contributors) to mathematical discussions.

Invitations to share, justify, or clarify thinking that position students as competent problem solvers, is relatively straight forward. However, I would warn that discursive positioning of this type should still be used with intentionality and caution. Keeping the social-emotional considerations of learning mentioned at the onset of this section front of mind, special emphasis should be placed on the fact that these discursive moves are *invitations* rather than requirements. Read your students. Sometimes cold-calling is warmed up on its own by the safe and exploratory tone of discussions. Other times, however, we can (and should) prepare students for sharing in a whole-class discussion by gathering responses ahead of time and sequencing them intentionally so that they build upon one another. This also gives us the opportunity to get students' permission to call on them to share their thinking during a whole-class setting. If they resist, consider restating or revoicing their ideas instead as a means of (a) getting their thinking into the whole-group space and (b) positioning them as valid contributors in the community and developing their agency. Some further examples of this second category of discursive positioning moves include eliciting responses based on prior conversations (e.g., "Bella, you and I were talking earlier about this. . . . Can you share how you were thinking about the problem? It was really interesting."), requesting for someone else to justify or clarify an idea, thus situating that person as a valid justifier or clarifier (e.g., "Muna, can you help us think about this? It's still a new concept for most of us. . . . What exactly is a *trapezoid*? How do you think about it?"), and invitations to think about someone else's thinking (e.g., "Somebody from last period disagreed with this result. Yonas, what about this do you think they were disagreeing with?").

Invitations for peers to respond to a student's idea in ways that position the idea as important or mathematically justified, is an excellent method of elevating a student's ideas and explanations—and thus the student. Importantly, we need to recognize the opportunity to elevate all discursive resources, such as noises, gestures, and drawings, in an effort to endorse multiple means and modalities of communication. Students are often able to conceptualize something internally before being able to formally communicate it or put it into words. When asked about the characteristics of a circle, students might draw a circle in the air with their finger and give some examples of things they know are round. Or, when asked how an exponential function appears to be behaving, students might make an upward curving hand motion and swooshing sound to illustrate an increasing growth rate or a function that is "speeding up." This thinking is individualized, novel, and deserves elevation if we are to promote mathematical ownership. We must develop a comfort, as teachers, amplifying *thinking* before attempting to formalize communication around discipline-specific conventions. Simply put, let's explore first and formalize later. In this sense, discursive positioning helps us center early learning around our learners.

Now that we have established multiple means of establishing a whole-class space that promotes mathematical ownership for all of our learners, how can we help ensure that these norms, behaviors, and practices translate to small-group interactions? In the next chapter, we will explore a number of practices and strategies that can assist with such a task.

CHAPTER 2 REFLECTION QUESTIONS

- What are some social or sociomathematical norms that students seem to arrive to your classroom expecting at the start of the year? What might this imply about their beliefs about mathematics?

- Are there social or sociomathematical norms that you try to develop in your classroom each year? Are there any that seem to develop on their own?

- What are your beliefs around the level of choice present in mathematics? Is it possible to overpromote choice in the subject?

- What ideas are new to you in the discursive positioning section? Are there any discursive positioning moves that you use regularly either with or without thinking about them?

Reinforcing Ownership Through Structured Peer Interactions and Collaboration

3

We're Doing It

We know it when we see it. The room is abuzz with the sounds of intent discussions, students are leaned in to each other's work, and we teachers are basically being ignored unless we are needed for arbitration or clarification. Whether it is a student explaining to his peers why $x = y^2$ is not a function of x by drawing a graph without prompting, a group debating the equivalence of reciprocal multiplication facts using counters arranged in arrays, students collectively coaching one another through the use of different addition strategies, or teams discussing the reasonableness of their mathematical solutions in a context, few things are as exciting to math teachers as excellent collaboration in the classroom. Much like those beautiful "ah-hah!" moments we are privileged enough to witness manifest in our students, these are the moments that give us pause and reflect on the fact that, *this time*, we really got it *right*. And while it is valuable to congratulate ourselves for getting our students to this space (for *our* senses of agency and efficacy are also important), the real power of reflection comes from working to determine *how* we helped create the conditions for such collaboration and *why* students are engaging in such a way. I aim to address these reflective questions in this chapter. Rather than just putting students together and hoping it all works out, let's backward plan collaborative tasks with this successful end in mind.

CHAPTER 3

Learning Intention:

I am learning how to reinforce student ownership of mathematics by structuring peer interactions and collaboration.

Success Criteria:

☐ I can elaborate on the value of collaborative learning in the mathematics classroom.

☐ I can strategically create heterogeneous groups that set the stage for collaboration.

☐ I can prime students for productive group work by thoughtfully launching tasks.

☐ I can emphasize choice and ownership in mathematics by creating opportunities for peer-assisted reflection.

☐ I can plan for collaborative protocols that suit the social and learning needs of my students.

Why Focus on Collaborative Learning?

If sitting down and listening to an expert propagate was all that we needed in order to learn mathematics then there would have been little need for the field of mathematics education over the last century—nothing would have needed changing! In fact, if this were the case, then the emergence of Khan Academy and other video-based virtual learning platforms would have made math teachers obsolete for the last decade or more. But *we're still here*. Why? To be clear, this is not an attempt to downplay these 21st-century learning tools, as they have incredible value, power, and potential to enhance student learning. Rather, I aim to situate them as exactly that—*tools*, and not replacements. Learning is personal; learning is custom; learning is human; learning is social. Teachers play *the* pivotal role in personalizing each student's education.

> Learning is personal; learning is custom; learning is human; learning is social.

Collaborating with others affords students opportunities to engage in mathematical content in ways that simply cannot otherwise occur. One such example is the opportunity to authentically use academic and content-specific language that, unfortunately, just doesn't seem to present itself as frequently in the lunch room. In other words, if we don't provide students the opportunity to actually *talk* about congruence, place value, rates of change, limits, and equivalence in our classrooms, these opportunities might not present themselves elsewhere. And to put it plainly, it is unrealistic to expect students to develop proficiency in the language of mathematics without allowing them the chance to *use* it in class.

CHAPTER 3 SUCCESS CRITERIA CALLOUT:

☐ I can elaborate on the value of collaborative learning in the mathematics classroom.

Setting up collaborative spaces in our classroom also serves as a permission slip to hold a high academic bar for our students. Genuine collaboration provides access to challenging content, to the Standards for Mathematical Practice, and to the Strands of Mathematical Proficiency. This process of mutual support is known as positive interdependence in the research world, whereby students scaffold one another into their *zones of proximal development*. Let's review. Vygotsky's (1978) zone of proximal development (see Figure 3.1) takes into account the dynamic nature of learning in that it aims to measure not only what a learner has already learned but also what they are capable of learning with expert guidance. This space between the *learned* and the *learnable* is what is formally deemed the zone of proximal development. The use of the adjective *proximal* is intended to describe the upcoming development that is within a student's reach, given the proper assistance. In other words, a learner's zone of proximal development contains what

Positive interdependence: the supportive process whereby students scaffold one another into their zones of proximal development through collaborative learning.

Figure 3.1   Zone of Proximal Development

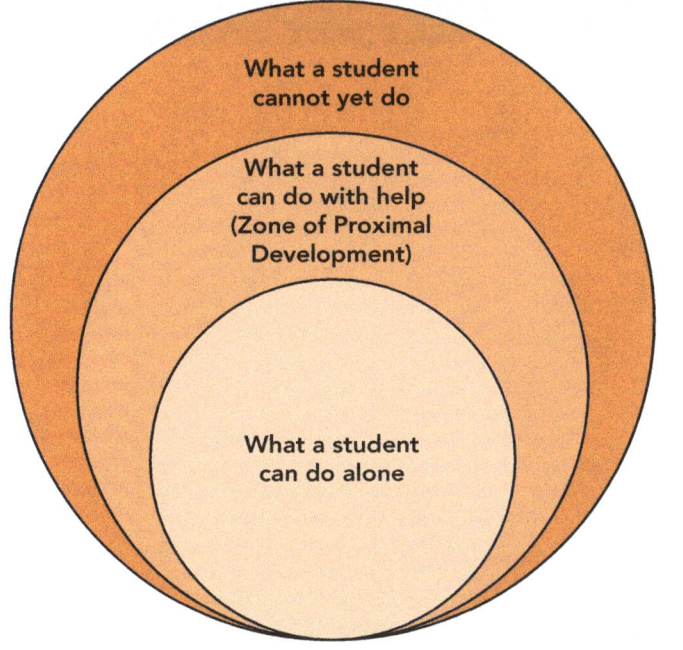

a student is ready to learn. Again, however, it should be emphasized that access to the contents within one's zone of proximal development must be mediated by "more capable peers" (Bunce, 2003; Moll, 1990; Vygotsky, 1978). Hence, we must provide collaborative opportunities if our students are to maximize their learning potential in our classrooms.

One of the primary ways through which we have been charged to ensure this rigorous mathematical experience for students is by engaging them in mathematical practices or processes. Take a moment and complete this next task (Table 3.1) before continuing with the reading. Look at the following commonly adopted mathematical practice standards and rank each on a scale of 0 to 3 regarding collaboration as follows (if your region uses a different set of practice or process standards, consider working through this process with those):

- **3**: Absolutely REQUIRES collaboration in order to access or exercise
- **2**: Can be *enhanced* by collaboration
- **1:** Would not be helped or hurt by collaboration
- **0:** Collaboration would detract from or hinder this practice or process

Table 3.1   Standards for Mathematical Practice Survey

STANDARDS FOR MATHEMATICAL PRACTICE		
STANDARD	**RANKING (0–3)**	**RATIONALE**
1. Make sense of problems and persevere in solving them.		
2. Reason abstractly and quantitatively.		
3. Construct viable arguments and critique the reasoning of others.		
4. Model with mathematics.		
5. Use appropriate tools strategically.		
6. Attend to precision.		
7. Look for and make use of structure.		
8. Look for and express regularity in repeated reasoning.		

 Available for download at **https://companion.corwin.com/courses/whosemathisit**

The hope is that this task provided a reflective moment on the fact that *most* of what students should be engaging in in our classrooms can (and should) be complemented by collaboration. What's more, certain practices can only be accessed through collaboration. Take Standard for Mathematical Practice 3, for example. How exactly does one *construct viable arguments and critique the reasoning of others* without others? And who is to be the audience of these arguments if students work alone? Are they writing letters? How long will that take to get feedback on the *viability* of their work? Clearly, for authentic engagement around that mathematical practice to occur, students need to be in collaborative spaces. Standard for Mathematical Practice 6, *Attend to precision*, addresses the need for precise language and notation in mathematics as to guarantee authors and speakers full control over the intended meaning of their messages. When we *attend to precision* with our communication, we don't leave the meaning of our work up to interpretation by audiences. Frankly, this is a pretty meaningless practice without *audiences*, or, others with whom to communicate.

There is an argument for collaboration enhancing each of these mathematical practices and processes—most of which are born from the space of creating authentic experiences for our students. This is not to say, however, that students should *always* be collaborating and working together in the mathematics classroom. Students absolutely need the opportunity to internalize these practices in order to truly claim ownership of mathematics. In this light, collaboration itself can act as a scaffold toward self-regulation and true independence, which we will cover in the next chapter. For now, let's talk about where to get started with collaboration in the math classroom: student groupings.

Priming Students for PRODUCTIVE Group Work

For years I have been beginning presentations to other educators about collaboration in mathematics by assigning them the following task (Figure 3.2) to complete in table groups. Note the rules.

I assign this task cold, only frontloading the instructions and rules. What I have observed across the country is fascinating. Some groups thrive, while others struggle. In some groups, natural leaders emerge and encourage others to engage, making sure no decisions are made until everyone is on board. In other groups, a single participant takes over and turns this into an individual task. In others still, certain members who don't feel as confident about their mathematical

Figure 3.2   Task-Order Matters

$\frac{3}{8}$ $\sqrt{2}$ $2\sqrt{3}$ -1 π 3 $-\sqrt{2}$ 3.7 $\frac{8}{3}$ $-\sqrt{81}$ $\frac{\sqrt{3}}{2}$	**Task-Order Matters** a. Arrange these numbers from *least* to *greatest* without a calculator. b. Place these numbers on a number line. Rules: 1 pencil and 1 paper per group

backgrounds lean away from the table and turn to technology to help them *wait it out* (I can't tell you how many times I have heard an administrator back away and say something along the lines of, "You all go ahead, I'm an administrator not a math person!"). Does any of this sound familiar? Much of our adult behaviors in social, collaborative spaces mirrors that of our students (or, perhaps it is the other way around?). Generating this sense of empathy for students is the purpose of this task. Now, consider these two reflection questions:

1. What are some of the skills you used to collaborate effectively with your groups?

2. What were some of the challenges?

CHAPTER 3 SUCCESS CRITERIA CALLOUT:

☐ I can strategically create heterogeneous groups that set the stage for collaboration.

For collaboration to be productive it needs two key ingredients: (1) team members need to be able to communicate effectively with one another, and (2) team members need to understand and have access to the task at hand.

Productive collaboration begins with creating groups of students who have access to both the content and to each another. The first half of this statement is why research suggests heterogeneous groups have a much higher impact on learning than long-term homogeneous groups (Hattie, 2009; 2023). If students are heterogeneously mixed based on current ability, there is a greater likelihood that at least one member from each group has direct access to the content at hand. The second half of the statement, however, that students must have access to each other, is why I don't suggest completely random groupings *or* groupings based on averaging (i.e., the lowest and highest perform-ing students on a recent assessment in the same group). Students typically work best together when they are academically *near* enough to be able to engage in meaningful, mutually supportive collabo-ration. If students are, at present, too far apart academically, they often withdraw into perceived roles of "explainer and listener," or worse yet, "worker and copier."

The sweet spot is being *strategically* heterogeneous with our student groups. Enter *alternate group ranking*—a method of creating a start-ing point for heterogeneous student groups based on recent assess-ment data. Alternate group ranking begins with fresh data—we need to know where our students' *current* performance lies. From there, we can rank and order our students based on this present data and group them using a few different methods, each of which is illus-trated in Tables 3.2 and 3.3 using an arbitrary case of 27 students in a class. In each case, the goal was to make as many groups of four as possible. Because we recognize the realities of teaching, we wanted to consider cases where our total number of students is not divisible by four.

> Students work best together when they are academically *near* enough to be able to engage in meaningful, mutually supportive collaboration.

Table 3.2 Alternate Group Ranking: Method 1

GROUP 1	GROUP 2	GROUP 3	GROUP 4	GROUP 5	GROUP 6	GROUP 7
Student 1	Student 2	Student 3	Student 4	Student 5	Student 6	Student 7
Student 8	Student 9	Student 10	Student 11	Student 12	Student 13	Student 14
Student 15	Student 16	Student 17	Student 18	Student 19	Student 20	Student 21
Student 22	Student 23	Student 24	Student 25	Student 26	Student 27	

Table 3.3 Alternate Group Ranking: Method 2

GROUP 1	GROUP 2	GROUP 3	GROUP 4	GROUP 5	GROUP 6	GROUP 7
Student 1	Student 3	Student 5	Student 7	Student 9	Student 11	Student 13
Student 2	Student 4	Student 6	Student 8	Student 10	Student 12	Student 14
Student 15	Student 17	Student 19	Student 21	Student 23	Student 25	Student 27
Student 16	Student 18	Student 20	Student 22	Student 24	Student 26	

In this first method, the top seven ranked students are sorted into seven different groups, acting as anchors for each. Then, the next seven ranked students are placed in the same seven groups so that Rank 8 is with Rank 1, Rank 9 is with Rank 2, and so on. This process is repeated until there are six groups of four students and one group of three students, as seen in the figure. This starting point works to address a few goals. The first is that every group has a student who recently achieved in the upper quartile—so direct access to the content is more likely across all groups. The second is that every student has another group member who scored within one quartile of them— so each student is grouped with somebody else academically *near* to them. Note the strategic attempt to address the preconditions of productive collaboration: (1) team members need to be able to communicate effectively with one another, and (2) team members need to understand and have access to the task at hand.

Now, to be clear, each of these methods of alternate group ranking simply create *starting points* for student groups based on quantitative achievement data. The next step is to apply a qualitative human lens and actually *look* at who landed where. Are there pairs of students that ended up in the same group that either haven't worked well together in the past or give you concern (overly social friends, students still resolving conflict, etc.)? Are there other students that you think *should* be grouped together that didn't end up in the same group (students who support one another with language development, previously recognized strong-pairings, etc.)? Finally, what is the general social makeup of each group? For example, is there a group of students who tend to be more introverted that might need a stronger conversationalist to help them work together? If any of the former concerns are present, try to address them by making lateral moves with students within quartiles. For instance, if Student 18 tends to support Student 9 with language acquisition, then swap Student 18 into Group 2 and Student 16 into Group 4. Additionally, if you notice that everyone in Group 3 tends to be a little quieter in class, and those in Group 7 participate more outwardly in group discussions, then consider a lateral move between members of each group. The point here is that alternate group ranking creates a mold, but the teacher can (and should) still put on the finishing touches.

This second method of alternate group ranking (Table 3.3) is one that I tend to recommend for younger students, as it acts as an additional social scaffold to aid conversation. In this method, you still gather fresh assessment data and rank students accordingly, but the sorting is different. Now, students are sorted as immediate pairs and placed into each group. So, Rank 1 and Rank 2 are placed into Group 1, Rank 3 and Rank 4 are placed into Group 2, and so on as seen in the figure. In this grouping method, the academic *nearness* between peers is drastically closer for pairs within groups. The idea here is that conversations around the content can occur more naturally and require less effort in each of these partnerships. When applying the same qualitative human lens to this starting point and making adjustments, keep the scaffolding goal of this academic nearness in mind. Sometimes, it can be helpful to zoom out and consider whole-group composition and shift pairs of students rather than individuals. For instance, you might look at Group 7 and be concerned that they might need more direct access to the content based on their members' overall recent performance (and the fact that they are only a group of three). Perhaps an adjustment of moving Student 9 and Student 10 to Group 7 and Student 13 and Student 14 to Group 5 would address this? Clearly, this is where knowing your students comes into play—as quantitative methods can only offer us a starting point for groupings.

CHAPTER 3 SUCCESS CRITERIA CALLOUT:

☐ I can prime students for productive group work by thoughtfully launching tasks.

In addition to designing groups that increase the likelihood of collaboration, we need to ensure that students have access to the tasks we assign them—for in the absence of access, no rigor exists. Priming students for success begins with considering what barriers might potentially exist that could block their access. This requires us to both know our students and to analyze our tasks. It might not be fair, for instance, to assume that students have a full understanding of all of the language embedded in the tasks we assign. They could have gaps in vocabulary (Do they know what all of the words *mean*?) or in understanding the language structure (Do they know what they are being asked to do?). Additionally, many rich tasks are embedded in real-world contexts so that mathematization and engagement in mathematical practices can be experienced in authentic and relevant ways. Contexts, however, could be partially or entirely unfamiliar to our students based on their life experiences and exposure, thus hindering access to the task.

In the absence of access, no rigor exists.

All of this is to simply say that rich tasks do not teach themselves. Teachers have a responsibility to launch tasks in a manner that maximizes access without reducing the rigor. Priming students for problem-solving is necessary and even prerequisite to productive engagement and opportunities to learn (Jackson et al., 2013), so how much time should we dedicate to it? And what are the factors worth considering when answering this question? We can think about priming students on a spectrum that ranges from doing nothing and launching cold on one extreme, to a full-day primer lesson on the other extreme (see Figure 3.3). Catalysts for sliding on this scale are considerations such as grade level, task complexity, prior exposure to the mathematics within the task, contextual familiarity, classroom culture and time of the year, as well as others. Here are some suggestions.

Figure 3.3   A Spectrum of Approaches to Launching Tasks

Alternative Forms of Media

If we anticipate that language, contextual features, or mathematical concepts could be relatively foreign to our students, we can consider using alternative forms of media to share information. Imagine being assigned a task about zip-lining or roof trusses if you have never been zip-lining or don't know what a roof truss is. Videos, pictures, and even constructed models of each could do a great deal to bring these contexts to life for students and situate them in a problem-solving space. Further, based on the depth of the tasks, there could exist some context-specific vocabulary that needs to be clearly developed and defined for students. Alternative forms of media make calling out and labeling these context-specific terms a much simpler task. We can even use alternative forms of media to scaffold access to mathematical concepts. Consider the examples in Tables 3.4 and 3.5.

Table 3.4   Scaffolding Example 1

MATHEMATICAL CONCEPT:	FRACTIONS
Media Suggestion	Video of someone measuring ingredients using a measuring cup for a recipe Physical measuring cups for students to examine
Discussion	Have students who have made food using measuring cups share their experiences.

Table 3.5   Scaffolding Example 2

MATHEMATICAL CONCEPT:	ANGULAR AND LINEAR VELOCITY
Media Suggestion	Video of merry-go-rounds Objects attached to different lengths of string to demonstrate spinning with different radii
Discussion	Have students who have ridden a merry-go-round or other similarly spinning rides share their experiences.

Close Reading

Another means of scaffolding linguistic and contextual access to a task is through the close reading strategy. Close reading is a text comprehension strategy whereby the teacher directs students to engage in multiple readings of a text with specific lenses for each reading. Teachers provide students with these specific lenses by asking text-dependent questions prior to each round of reading and then facilitating subsequent whole-class discussions. This back-and-forth flow of reading and whole-class discussions is intended to prime all students for sensemaking. The general logic here is that students can first develop a broad understanding of the context by responding to very general and literal text-dependent questions, then start to decipher the purpose of the text and what it is that they are actually tasked to do by answering questions that explore the text's structure, and then ultimately investigate the inferential meaning and mathematical implications of the text so that they might start crafting a problem-solving plan. Figure 3.4 demonstrates this flow (Fisher et al., 2015).

Figure 3.4   Progression of Text-Dependent Questions

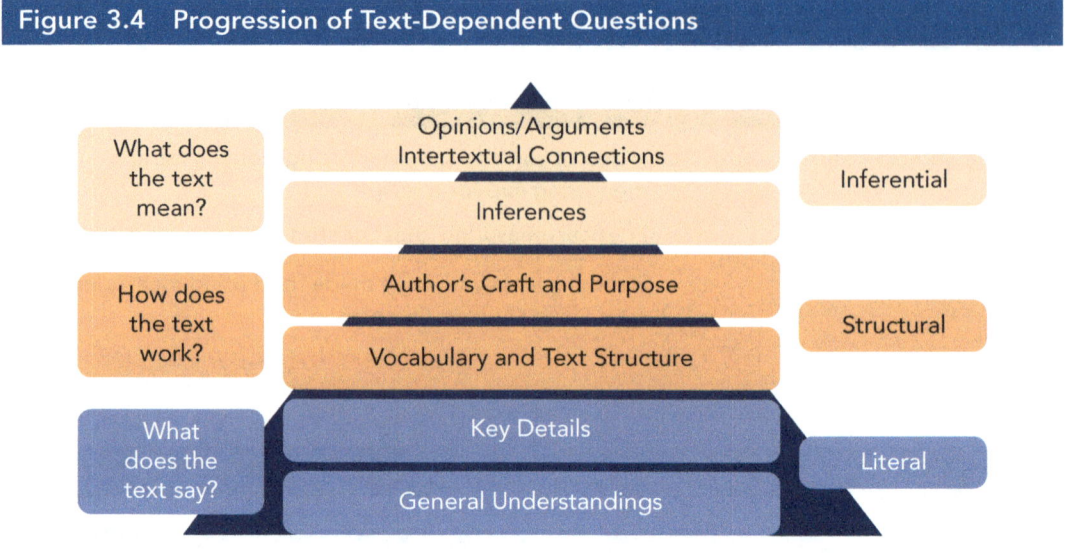

Source: Adapted from Fisher and Frey (2015).

When planning for a close reading primer lesson, teachers should begin by actually completing the task that they are assigning students. While doing so, they should note how they are using the following in order to *make sense* of the task:

- **Content Language** (Language that is domain specific, has a static definition/meaning. e.g., *rectangle, place value, decimal, polynomial, denominator*)

- **Academic Language** (Language that is used across domains, often with different meanings. e.g., *area, angle, function, derivative, domain*)

- **Context of the Task** (How the context contributes to overall understanding. e.g., *Tides*—result in different water levels at different times; change periodically; caused by the Moon's gravitational pull. *Buying a car*—financing can require a down payment and equal monthly payments; interest is charged and compounded monthly; paying cash avoids interest charges.)

- **Connections Across the Text**

Consider using the reproducible Deconstructing the Text template seen as Figure 3.5 as a space to capture this information. A full-page PDF version can be downloaded on the book's companion website: **https://companion.corwin.com/courses/whosemathisit**

Figure 3.5   Deconstructing the Text Template

Text-dependent Discussion Questions–Deconstructing the Text

Name of Text: _____

Content Language: (Domain specific, static definition/ meaning; e.g., decimal, *polynomial, denominator.*)	**Academic Language:** (Used across domains, often with different meanings; e.g., *function, derivative, domain.*)

Contextual Inferences or Implications: (How is the context contributing to your overall understanding? e.g., *Tides*—result in different water levels at different times; change periodically; caused by the Moon's gravitational pull. *Buying a car*—financing can require a down payment and equal monthly payments; interest is charged and compounded monthly; paying cash avoids interest charges.)

This raw, sensemaking information can then be used to inform the teacher's creation of text-dependent questions across all questioning tiers. Essentially, we aim to write questions that guide students to notice features that helped *us* make sense of the task. The reproducible *Text-dependent Discussion Questions* form in Figure 3.6 can be used to help ensure you are sequencing text-dependent questions from each tier. A full-page PDF version can be downloaded on the book's companion website: **https://companion.corwin.com/courses/whosemathisit**

Figure 3.6   Text-Dependent Discussion Questions Template

Text-dependent Discussion Questions—Deconstructing the Text

Name of Text: _____

CATEGORY	WHAT QUESTIONS COULD YOU ASK STUDENTS HERE?
What does the text say? **LITERAL** (General understanding and key detail questions about the content.)	
How does the text work? **STRUCTURAL** (Vocabulary and text structure questions to bridge explicit and inferential meaning—including mathematical meaning.)	
What does the text mean? **INFERENTIAL** (Going deeper to mine author's purpose, inferences across the text, locating/connecting meaning across multiple texts.)	
What culminating question or task follows from this? What does the text inspire?	

Source: Fisher and Frey (2023).

 Available for download at **https://companion.corwin.com/courses/whosemathisit**

Go ahead and try this process, using the provided forms, with the upcoming task involving quadratic functions in a real-world context. Following the task (Table 3.6), you will find a sample text-deconstruction (Table 3.7), along with potential text-dependent questions (Table 3.8).

Table 3.6 Sample Mathematical Rich Task for Close Reading

Fireworks: Performance Task

Safety First . . . then BOOM!

Source: iStock.com/Corri Seizinger

<u>Scenario</u>

Your team has been hired to ensure the safety of a city's upcoming firework show. Your goal is to write a safety proposal after reviewing each source of research. You must document how you came to your recommendation (show your work and explain your thinking, don't just write an answer), and note how your proposal changes as you review more sources.

Source 1. A firework contains a time-delay fuse that burns as the firework soars upward. The length of this fuse must be made so that the firework does not explode too close to the ground. In order to be safe for those on the ground, a firework must explode at a height of at least 450 feet. The height (in feet) of a firework t seconds after it is launched is modeled by the function $h(t) = 3 + 180t - 16t^2$. The average fuse burns at a rate of 0.37 inches per second.

First Draft Proposal:

What range of fuse lengths should be considered safe? Support your claim with evidence from the text and mathematical reasoning.

Table 3.7   Sample Completed Deconstructing the Text

CONTENT LANGUAGE: (DOMAIN SPECIFIC, STATIC DEFINITION/ MEANING. E.G., *DECIMAL, POLYNOMIAL, DENOMINATOR*.)	ACADEMIC LANGUAGE: (USED ACROSS DOMAINS, OFTEN WITH DIFFERENT MEANINGS. E.G., *FUNCTION, DERIVATIVE, DOMAIN*.)
Firework—shoots into the air and then explodes. **Time-delay fuse**—a **fuse** is something that ignites an explosive device. **Time-delay** means that it won't explode immediately. This implies the user has some control over when to detonate. **Length**—distance from one end of the fuse to the other when taut. **At least 450 feet**—safe to explode at height ≥ 450. Also implies some time must pass. **0.37 inches per second**—Every second that passes 0.37 inches will burn. This seems somewhat slow because it will take almost 3 seconds for a full inch to burn.	**Soars upward**—colorful way of saying the firework shoots vertically into the sky before exploding. **Launched**—the ignition process that shoots the firework up into the sky. **Modeled**—in this context, this means *mathematically modeled*, whereby a function is used to predict the firework's height after a certain amount of time. **Average**—though not explicitly stated, it is safe to assume this represents the *mean*. **Rate**—in this context, a ratio comparing two changing quantities (distance and time). **Range of fuse lengths**—the set of distances a fuse could burn to allow the firework to travel at least 450 feet. This implies multiple answers.

Contextual Inferences or Implications: (How is the context contributing to your overall understanding? e.g., *Tides*—result in different water levels at different times; change periodically; caused by the Moon's gravitational pull. *Buying a car*—financing can require a down payment and equal monthly payments; interest is charged and compounded monthly; paying cash avoids interest charges.)

Fireworks begin as vertical projectiles. They are shot into the air and have gravity pulling them back down toward the ground.

SAFETY—This implies that if unexploded, they might reach 450 feet, go higher to their maximum, then start falling, reach 450 feet again, then continue falling. This is why they can be modeled by a **negative quadratic function** (upside down parabola). They must explode in that "sweet spot" in order to be safe.

Connection between fuse-length and height

Fuse-length——[determines]——> **Time in air** ——[determines]——> **Height of explosion**

Table 3.8 Sample Completed Text-Dependent Discussion Questions

CATEGORY	WHAT QUESTIONS COULD YOU ASK STUDENTS HERE?
What does the text say? **LITERAL** (General understanding and key detail questions about the content.)	• What is this text about? • What is our goal?/What are we supposed to ultimately produce? What are fireworks, and how do they work? Restate or explain the first sentence in your own words to your table. According to the text, what is the relationship between *safety* and *height*? Where are you getting that information? (elicit sourcing)
How does the text work? **STRUCTURAL** (Vocabulary and text structure questions to bridge explicit and inferential meaning— including mathematical meaning.)	• How might you describe a "time-delay fuse" to someone who doesn't know what it is? • What is the meaning of the word "modeled" in this context? • According to the text, how might we determine the height of the firework after a certain amount of time? (Formative follow-ups: What's the height of a firework after 1 sec, 2 sec, 5 sec? Are any of these heights safe for explosion?) • What does the sentence *The average fuse burns at a rate of 0.37 inches per second* mean in math terms? • FOCUSING: How could we use this information to determine how much fuse burns after a certain amount of time? • FUNNELING: Write a function, $F(t)$, that models the length (in inches) of a fuse that has burned after **t** seconds.
What does the text mean? **INFERENTIAL** (Going deeper to mine author's purpose, inferences across the text, locating/ connecting meaning across multiple texts.)	• Why do you think it is important to use a time-delay fuse in fireworks? • What is the connection between fuse length and height? Follow-up: If I told you a firework was in the air for 5 seconds before it exploded, could you tell me how high it went? Could you *also* tell me how long the fuse must have been? How can we generalize this?

What culminating question or task follows from this? What does the text inspire?

What should our safety proposal include? What type of evidence is appropriate to include? (Graphs, calculations, written explanations/annotations of work.)

What form should our answer take? SENTENCE FRAME: *Fuse-lengths between* _____ *and* _____ *should be considered safe because* _____.

Three Reads Protocol

Not all tasks are text heavy enough to require a full close reading lesson, especially at the elementary level. A less-robust alternative for lighter tasks/word problems is the Three Reads protocol (San Francisco Unified School District [SFUSD] Mathematics Department, 2015). Similar to close reading, the Three Reads protocol (see Table 3.9) funnels students from the broad to the specific by rereading a problem with multiple lenses. The first read is often led aloud by the teacher with a focus on the question, *What is this situation/problem about?* Again, much like close-reading, this first read is followed up with partner or group talk and a whole-class discussion. After consensus is built, the second read is either a choral or partner read with a focus on the question, *What are the quantities in the situation/problem?* This lens allows students to consider the units involved in the context and thus begin mathematizing the context. The third read—again either choral or partner-to-partner—has a focus on the question, *What mathematical questions can we ask about the situation/problem?* This question provides students the opportunity to start consolidating the more surface-level information they gained from the first two reads and start making mathematical inferences. Once the teacher is confident that students are sufficiently situated in a problem-solving space, they are directed to work on the task in their groups.

Revisiting and Reinforcing Social and Sociomathematical Norms

The work of shaping classroom culture is never complete. Priming students for success on collaborative tasks often spills over into the social demands of the task itself, as well as the task's mathematical expectations. Take a look back at the *Assessing Social Norms Teacher Questionnaire* in the last chapter. Many of those social learning intentions were written with the demands of specific tasks in mind. Take, for instance, the social learning intention, *To collaboratively and respectfully challenge the thinking of our peers.* This social learning intention can be elaborated during the launch of a task with a focus on crafting viable arguments and critiquing reasoning. Social learning intentions like these can set the tone and reinforce the social norm that we work together, we challenge each other for everyone's benefit (including by playing the devil's advocate), and we do so with respect. Further, consider the social learning intention, *To use what you know to make sense of a new concept.* Elaborating a social learning intention like this can prime students to project existing, securely held mathematical structures onto otherwise unfamiliar mathematical contexts. In this sense, activating

Three Reads Protocol Summary

Table 3.9 Three Reads Protocol Summary			
	WHAT THE TEACHER DOES	**KEY QUESTION**	**WHAT THE STUDENTS DO**
Preparation	• Identifies appropriate problem stem • Anticipates linguistic and mathematical challenges • Creates visuals to support understanding		
1st Read	• Shows visuals • Orally reads the "story" (problem stem)	What is this situation/problem about?	• Sit with a partner • Listen to the "story" • Turn to partners to discuss the "story" in their own words • Say what they remember of the story
2nd Read	• Shows problem stem (for example, on overhead projector or poster) • Leads class in choral or partner read • Leads discussion of quantities and units	What are the quantities in the situation/problem?	• Read chorally with the class or with partners • Volunteer quantities and units they identify
3rd Read	• Asks partners to read with specific goal • Leads discussion of potential questions • Clarifies language as needed	What mathematical questions can we ask about the situation/problem?	• Read one more time with partners • Brainstorm with partners several questions that could be asked using the problem stem • Volunteer questions

Source: San Francisco Unified School District (2024).

social and sociomathematical norms can act much like a prompt for students in that it brings relevant dispositions front-of-mind as they prepare to approach a collaborative task.

Peer-Assisted Reflection (PAR)

Not all collaboration in the mathematics classroom comes in the form of students in small groups working collectively on the same task. Students can also collaborate by providing one another with feedback on their individual work, which aids in developing individual mathematical ownership. Much like small-group collaboration, however, this type of student-to-student feedback should be structured in order to ensure productive participation and maximize student benefit. Peer-Assisted Reflection (PAR) is one such tool to structure peer-to-peer feedback as well as foster individual reflection and ownership of mathematics (Reinholz, 2015).

CHAPTER 3 SUCCESS CRITERIA CALLOUT:

☐ I can emphasize choice and ownership in mathematics by creating opportunities for peer-assisted reflection.

PAR is a collaborative protocol that promotes conceptual understanding through an iterative problem-solving approach (Reinholz, 2015). PAR requires students to

1. Work on meaningful problems

2. Reflect on their own work

3. Analyze a peer's work and exchange feedback

4. Revise their work based on insights gained throughout this cycle

The feedback process begins with approximately 5 minutes of silent reading and annotating of one another's work. Then, the partners debrief their work and feedback. As homework, students revisit and revise their work based on their peer feedback and submit a final solution to their teacher for grading or scoring purposes.

In both phases of Reinholz's study (2015), students engaging in PAR outperformed their control group peers on every single course exam and passed the course at a 13% greater rate and a 23%

greater rate respectively. Clearly, the PAR process has a positive effect on learning.

The focus of PAR is on developing students' ability to explain their thinking in an effort to round out their conceptual understanding. When students meet in pairs to review one another's work, these conversations need to be productive and targeted. Students should be conditioned into the PAR process by engaging in whole-class PAR sample analyses, where the teacher questions the acceptability of various lines of sample student work and subsequent explanations. In other words, students should be taught how to analyze work and provide one another with feedback. This is also an excellent opportunity to develop and renegotiate the social and sociomathematical norms discussed in the last chapter. Further, in developing the PAR protocol prior to this study, Reinholz (2015) noted that students would often link up with friends for feedback and the resulting conversations were answer-oriented (rather than explanation- and process-oriented). This is why it is suggested to either randomly assign partners or strategically pair students toward positive interdependence.

Note that it is important to have a plan for when neither student in a review pair has an understanding of the week's problem. Rather than limiting students to a discussion about how neither of them understand the problem, consider alternate ways to scaffold the problem, provide reteaching, or give them other tools to help solve the problem.

Reinholz (2015) reports the benefits of PAR through the acronym IDEA—iteration, discussion, explanation, and alternatives. PAR encourages students to approach mathematics iteratively (something natural in the field) by compelling them to revisit their work and engage in a draft-solution process (something surprisingly novel in mathematics but not at all foreign to the writing process in English language arts). PAR also promotes discussion by structuring weekly feedback sessions with peers. Additionally, PAR emphasizes explanation and sets the standard of knowing as one's ability to explain their thinking to someone new rather than exclusively seeking the correct answer. Finally, PAR exposes students to alternative ideas, methods, thinking, and problem-solving paths by virtue of its cross-pollinating nature (Reinholz, 2015). Figure 3.7 shows two PAR examples.

Figure 3.7a   PAR Example 1

Math 3 Peer-Assisted Reflection – 1.4   Name: Period: Date:

Write your solution in the left column. The right column is used for annotations. If you provide feedback to your peer, you will annotate their solution. After class, you will annotate your own solution as well. In your submission, use the annotation column to explain how you did (or didn't) respond to peer feedback.

PAR 1.4 – Prediction Equations

Success Criteria

[ ] Use scatter plots and describe correlations. **(A, B, C)**

[ ] Model data using lines of fit and their prediction equations. **(D, E)**

The table below shows the study times and test scores for a number of students. Use the table to complete the following:

Study Time (min)	11	14	20	27	34	39	41	45
Test Score	58	62	67	71	71	75	72	77

A) Create a scatter plot.

B) Choose two points to draw a line of fit. EXPLAIN why you chose the points you did.

C) Describe the correlation. JUSTIFY your description.

D) Write a prediction equation.

E) Predict the test score of someone who studied for 1 hour. How confident are you in this prediction? Does this seem accurate? Why or why not?

Reviewed by: _____

Rate your peer's mastery of the success criterion (this is the *last* thing you do):

[ ] Use scatter plots and prediction equations.

0 – DO NOT CHECK THAT BOX	1 – ALMOST CHECK THAT BOX	2 – CHECK THAT BOX
Many mathematical errors and/or incomplete or unclear annotations.	Few mathematical errors and/or somewhat incomplete or unclear annotations.	No mathematical errors and perfectly complete and clear annotations.

[ ] Model data using lines of regression.

0 – DO NOT check that box	1 – ALMOST check that box	2 – CHECK that box

<u>DRAFT SOLUTION</u>

<u>ANNOTATIONS</u> (Author's AND Peer's)

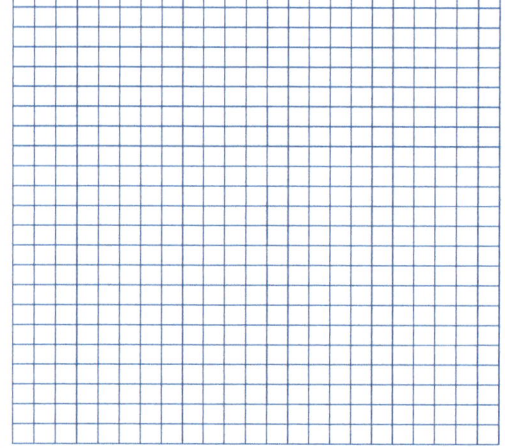

 online resources ↖

The full PAR example is available for download at **https://companion.corwin.com/courses/whosemathisit**

Figure 3.7b PAR Example 2

Unit 4 Peer-Assisted Reflection

Write your solution in the left column. The right column is used for notes. If you provide feedback to your partner, you will annotate their solution.

For homework, respond to your partner's feedback. Or, reflect on your solution if your partner did not have any feedback.

Success Criteria:

[ ] I can calculate data points in a table.

[ ] I can use data to graph the points on a coordinate plane.

[ ] I can analyze data and make conclusions, generalizations, or predictions.

Use the table and graph on the next page to complete the problem.

A cyclist rides at a constant speed of 20 kilometers per hour.

1) Complete the table to show the distance she travels after 1, 2, 3, and 4 hours of riding.

2) Graph the data on the coordinate plane.

3) Answer the following question: If the cyclist were to keep riding for 6 hours, how far would you expect her to go? Explain your reasoning.

Reviewed by: _____

Rate your partner's mastery of the success criterion. (This is the last thing you do.)

[ ] I can calculate data points in a table.

0	1	2
DO NOT check that box	ALMOST check that box	CHECK that box
Many math errors and/or incomplete or unclear work	Few math errors and/or somewhat incomplete or unclear work	No math errors and perfectly complete and clear work.

[ ] I can use data to graph the points on a coordinate plane.

0	1	2
DO NOT check that box	ALMOST check that box	CHECK that box

[ ] I can analyze data and make conclusions, generalizations, or predictions.

0	1	2
DO NOT check that box	ALMOST check that box	CHECK that box

Time (hr)	0			
Distance (km)	0			

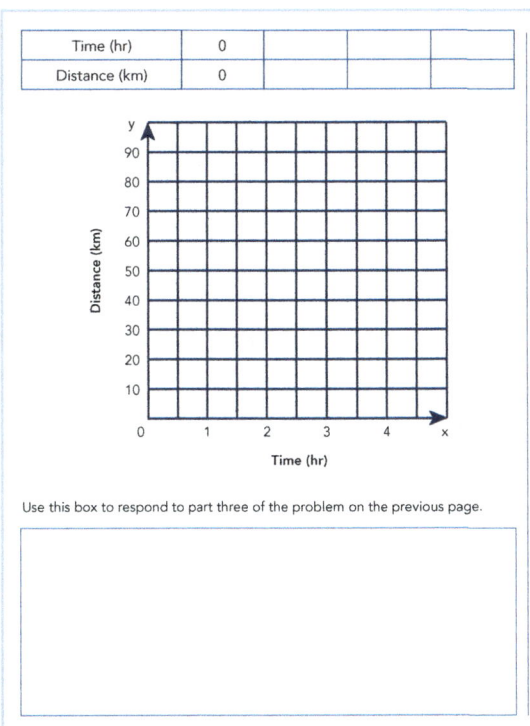

Use this box to respond to part three of the problem on the previous page.

Other Collaborative Protocols in Math

In this final section of the chapter, I will delineate a number of other collaborative protocols and strategies suitable for mathematics. This is, by no means, meant to be an exhaustive list. Rather, the goal here is to provide a repertoire of tried-and-true structures for various instructional situations. Not everything in mathematics is a text-heavy, context-embedded rich task, and thus, collaborative spaces should not always be structured for such. Sometimes, we just need ways to scaffold and motivate partner talk, for example. My hope is that this section can provide ideas to get readers started across the collaborative spectrum.

CHAPTER 3 SUCCESS CRITERIA CALLOUT:

☐ I can plan for collaborative protocols that suit the social and learning needs of my students.

Max and Min—Community Building

Sometimes we need to break the ice before we are able to engage in collaborative work with others with whom we are not overly familiar. Other times, we know our partners well but have not seen them for some time—so we absolutely have to check in before engaging in collaborative work. These are the social realities of being human. The Max and Min protocol serves to address these human needs by providing students a structured social space to engage in with their group members prior to engaging in academics. This is often valuable at the start of the school year or after students return from an extended break. At the start of a school day/class period, have students sit in small groups and provide time to discuss the question, *What's the best thing that happened over break?* And, completely optionally, *What's the worst?* It is important to emphasize the optional nature of the second question, as to not enforce oversharing that students might not be comfortable with. Providing students with about 30 seconds per group member is sufficient for this first phase. While students are discussing their time away from school, the teacher can walk the room and add social support where it might be needed. Look for students who aren't talking to each other, and help them start the conversation—make sure all group members are being included in conversations. After students have a chance to discuss in their groups, ask each group to decide on an *absolute max* and—optionally—an *absolute min* from what was shared in their group to share with the whole class. Facilitate a brief share-out from each group, limiting responses to one per group and roughly 45 seconds per group. This whole protocol should take less than 10 minutes. The purpose here is to get students *talking* before asking them to collaborate about mathematics. This protocol can help remove social barriers by building community between group members before they engage in the strenuous work of mathematical sensemaking.

Numbered Heads Together

When a lesson's design calls for a lot of back-and-forth between small-group interaction and whole-group discussion, maintaining individual student accountability can become more challenging. Further, hearing from a diverse array of voices during whole-group discussion is rarely something that comes standard issue with classes. I have read before that cold-calling is the answer, while reading elsewhere that cold-calling can be detrimental to student participation. How, then, can we keep all students leaned-in and engaged? Numbered heads together (Kagan & Kagan, 2021) is a protocol that infuses randomization with preparation in order to support motivation and quality of responses. Here's how it works: Students are asked to work on a small task as a group. Each group is uniquely numbered as is each student within each group. As students make progress toward a planned instructional pausing point, the teacher rolls a die or uses some other random number generator to determine which *student number* will be called on to share the group's thinking. For instance, if the teacher rolls a number 4, she might communicate to the class, "All right folks, it's a four! That means if you are Student 4 in your group, then you have the possibility to be called on in the next round to answer this question. Make sure you are prepared! Other group members—help your number 4s get prepared!" The students are then given a few more minutes to reengage with the task with the knowledge of who has the potential to be called on for whole-group sharing. Students who know that they might be called on are now extra motivated to synthesize their group's thinking so that they are prepared for sharing. This eliminates the risk of true cold-calling, while still allowing for randomization. In the final step of this strategy, after students have had the opportunity to reengage with the task, the teacher rolls or randomizes again in order to determine which *group number* will be called on to share. So, if the teacher rolls a 3 this time, then Student 4 (determined last round) from Group 3 will share out to the whole class. This strategy can be repeated throughout a lesson and used as the driving structure of the day. As counterintuitive as it may sound, an interesting and comforting sense of predictability can develop in our students from this planned unpredictability. Finally of note, each subsequent whole-group share-out is an excellent opportunity for teachers to engage in the discursive positioning moves, such as restating or revoicing, discussed at the end of the last chapter. In addition to holding individual students accountable, this protocol should facilitate our elevation of individual students and promotion of their sense of mathematical ownership.

Collaborative Sentence Frames

Collaboration, we should remember, is a skill. Learning to engage in productive discussions, sharing the air, and participating equitably are all aspects of collaboration that students could need support in developing. One manner of supporting students develop the skills of collaboration is through the use of collaborative sentence frames. These sentence frames provide a structural scaffold for the language of productive participation and can serve as a reminder for students of what it means to collaborate. These sentence frames can be designed to help students share their thinking, probe the thinking of others, ask follow-up questions, build consensus, and more. Table 3.10 includes a sample of these sentence frames that can be placed on table tents on each group's workspace or displayed as an anchor chart during highly collaborative lessons.

Table 3.10   Collaborative Sentence Starters

Collaborative Sentence Starters

I think the answer is _____.

My strategy was _____.

I figured it out by _____.

I think this is because _____.

I know that _____ because _____.

I don't think that will work because _____.

My answer is reasonable because _____.

I agree/disagree with _____ because _____.

I can prove my thinking by _____.

I'm still not sure about _____.

I'm wondering about _____.

Help me understand why this _____.

Collaborative Posters

Collaborative posters are an excellent means of promoting individual accountability on larger tasks. In this protocol, each group receives a large piece of poster or chart paper, and each group member is assigned a different color marker. Each member can only write on the poster in *their* color, as to clearly identify each member's contributions. This method tends to have a self-regulating effect, as it becomes abundantly clear when certain members are not contributing in proportion to others. At the end of the task, all members sign the poster in their color in order to note which contributions were theirs. This color-coded means of contribution also scaffolds the teacher's job when scanning the room. During the task, it remains clear which groups or group members might need more direct support.

Gallery Walks

Gallery walks can be coupled with the previous collaborative poster strategy—if suitable for the task—or used in tandem with other types of poster/creative/problem-solving tasks. After students produce some form of mathematical work that can be displayed on the wall (sticky chart paper is the best vehicle for this), groups rotate throughout the room, from display to display, leaving comments and feedback for each other on sticky notes, while also taking notes for themselves. This is an opportunity for students to see different problem-solving strategies, data-displays, or evaluate one another's mathematical claims; the gallery walk protocol affords many different uses. Figure 3.8 shows an example of student work from a gallery walk. In this task, students were to use data to make a claim about whether a professional athlete of their choice was getting better or worse at their sport. The students who created the poster shown in Figure 3.8 chose to investigate Michael Jordan's career statistics, including points per game, rebounds per game, assists per game, and field goal percentage, in order to make a claim about his performance over time. This data, its usage, and the claim itself would all be evaluated by other students during the gallery walk.

Figure 3.8　Sample Student Work

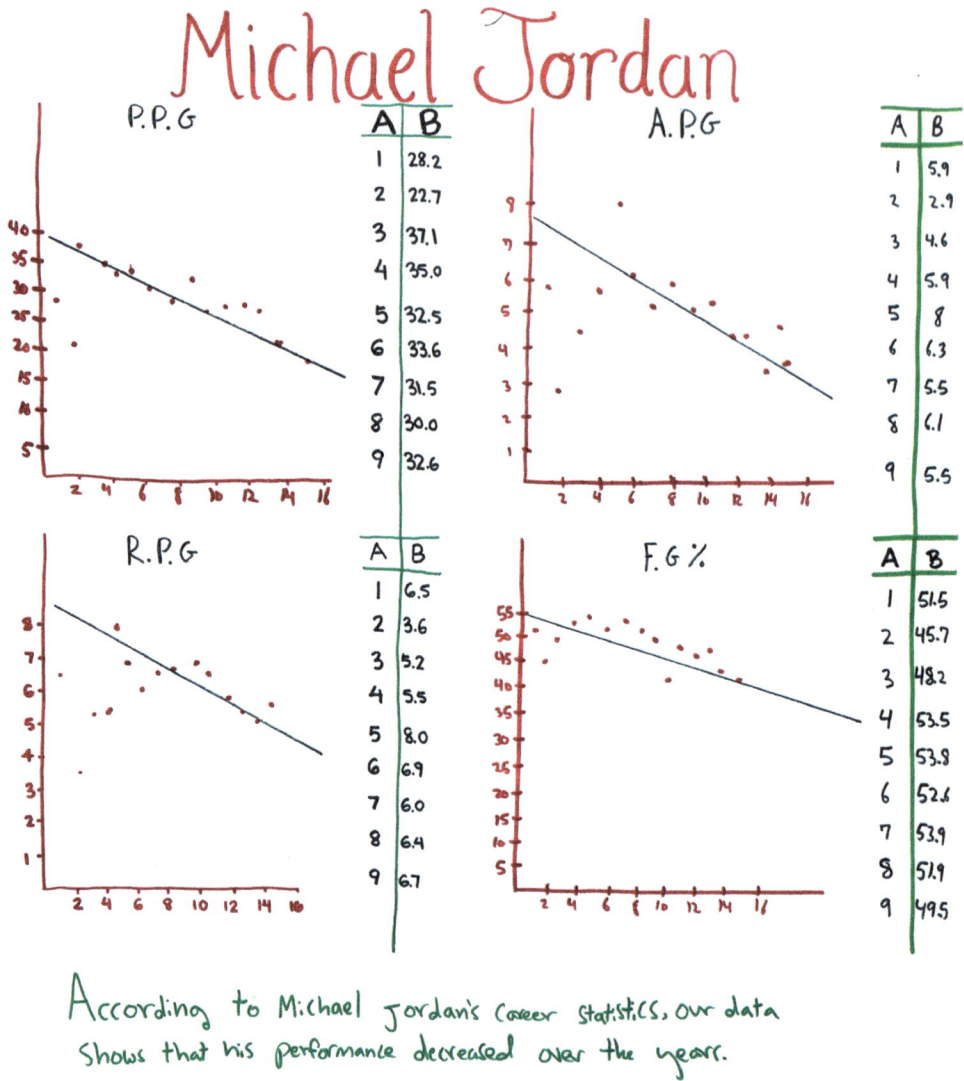

Michael Jordan

P.P.G

A	B
1	28.2
2	22.7
3	37.1
4	35.0
5	32.5
6	33.6
7	31.5
8	30.0
9	32.6

A.P.G

A	B
1	5.9
2	2.9
3	4.6
4	5.9
5	8
6	6.3
7	5.5
8	6.1
9	5.5

R.P.G

A	B
1	6.5
2	3.6
3	5.2
4	5.5
5	8.0
6	6.9
7	6.0
8	6.4
9	6.7

F.G%

A	B
1	51.5
2	45.7
3	48.2
4	53.5
5	53.8
6	52.6
7	53.9
8	51.9
9	49.5

According to Michael Jordan's career statistics, our data shows that his performance decreased over the years.

When considering which strategy to use for a given lesson, it is important to remember that protocols themselves are not self-important. Protocols are scaffolds—if our students don't need them then we shouldn't use them. We have to be ready to provide structure and support while also being aware of when we need to get out of the way. Because, again, our focus is on fostering *ownership* of mathematics, not rigid compliance. Speaking of mathematical ownership, now that we have discussed how to create a whole-class culture that promotes mathematical autonomy and ownership, and provided examples of how to help transfer that culture of ownership into collaborative spaces, how can we help to ensure that students maintain and internalize this sense of ownership at the individual level? In the next chapter we will work toward that exact end.

CHAPTER 3 REFLECTION QUESTIONS

- What are some examples of collaboration "going right" in your classroom? What were some of the conditions you think that might have led to that success?

- Which method of alternate group ranking do you think would be more beneficial for your students? What are some additional considerations you might have to take into account when making student groups?

- Overscaffolding reduces the mathematical rigor of tasks and robs students of the cognitive conflict we know to be prerequisite to learning. Language and context, however, cannot be overscaffolded. How might you determine *when* students have access to a task and are ready to begin working on their own?

- Some teachers have chosen to use PAR tasks in lieu of quizzes, while others have replaced traditional homework assignments with them. Others still have woven them into their existing assignment loads. How might you see these types of peer-assisted reflection tasks fitting into your instruction?

- What are some other collaborative protocols you have had success with? Are there contingencies you would place on any?

Promoting Metacognition Through Self-regulated Learning and Feedback

4

I'm Doing it

Mathematical ownership is not a timeshare operation. Students cannot successfully operate as partial owners, relying on the collective to make up for what they do not possess individually each time a problem needs solving. So, while the collaboration we foster, promote, and pursue is of the utmost importance on the path to mathematical ownership, it ultimately exists as a scaffold and vehicle toward *individual* ownership, which in turn lends itself toward greater collective feats. I urge you not to mistake this for a larger argument about the merits of individualism versus collectivism but rather to view it as a claim that specialization should happen at a deeper level of content. When it comes to K–12 mathematics, every learner is entitled to individual access to every standard. In this chapter, I aim to illuminate how to move students from saying *We're doing it* to *I'm doing it*.

In order to foster self-regulation in our students, we must first understand it. This is why the chapter begins with a deep dive into some of the research and existing literature on self-regulation and motivation. By breaking these broad concepts into a framework of subprocesses, developing our students as self-regulated learners can become more targeted, comprehensive, and manageable. From there, I will offer some classroom strategies to support this effort. In addition to creating opportunities for students to recognize and engage in self-regulation, we need to ensure that students know how to productively act on their own data. Hence, a large portion of this chapter will be dedicated to building independent study skills that are appropriate for mathematics. I hope to provide you with strategies for promoting individual mathematical ownership that are both practical and ready to implement, as well as grounded in the research.

CHAPTER 4

Learning Intention:

I am learning how to consolidate each student's ownership of mathematics by fostering their self-regulation of learning.

Success Criteria:

- ☐ I can articulate the motivation cycle and its components.
- ☐ I can delineate self-regulation into its three phases and describe each.
- ☐ I can build opportunities into my course for students to explicitly engage in self-regulation of learning.
- ☐ I can create opportunities for students to engage in self-assessment.
- ☐ I can articulate the difference between formative and summative assessments in order to ensure intentional and impactful implementation.
- ☐ I can identify multiple study skills that are appropriate for mathematics and teach my students to use them effectively.

Motivation and Self-Regulation

What does it mean to be self-regulated? Zimmerman (2002) states that self-regulation is not a skill but a self-directive process that involves self-awareness, self-motivation, behavioral skill, and knowledge of a skill, and thus situates it as something that everyone engages in to varying degrees. In fact, many of the differences between learning outcomes in students can be attributed to differences in their metacognition and self-regulation (Zimmerman, 2002). The argument here is that if students

CHAPTER 4 SUCCESS CRITERIA CALLOUT:

- ☐ I can articulate the motivation cycle and its components.

are more aware of their personal limitations, they will be better able to compensate. In other words, if they know what they need and they know where to find it, students are more likely to take corrective action. There is an air of student empowerment in this concept, reminiscent of the impact of teacher clarity discussed in an earlier chapter (Houser & Frymier, 2009).

Zimmerman (2002) further reports that, "Because of their superior motivation and adaptive learning methods, self-regulated students are not only more likely to succeed academically, but to view their futures optimistically" (p. 66). Again reminiscent of the previously reported research on teacher clarity, this might remind you of the impact that teacher clarity has on affective learning (Titsworth et al., 2015). And while he notes that beginners in any field or discipline run a heightened risk of low motivation, Zimmerman (2002) states that this can be thwarted by self-regulatory processes, such as self-monitoring, that allow learners to witness their progress more explicitly. This, in turn, positively affects their self-satisfaction and personal efficacy. In summary, it would appear that progress supports progress and success breeds success. This cycle of motivation is summarized in Figure 4.1.

Self-regulation: a self-directive process that involves self-awareness, self-motivation, behavioral skill, and knowledge of a skill, and thus something that everyone engages in to varying degrees

Figure 4.1 The Motivation Cycle

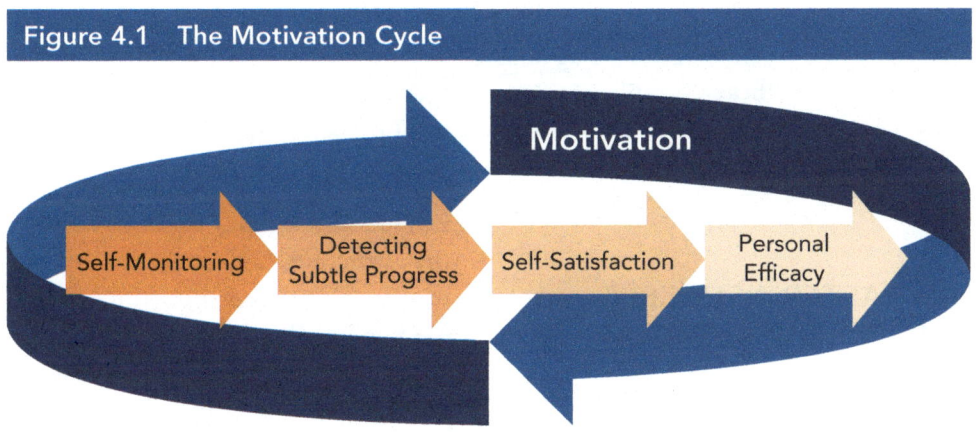

Fundamentally, these are the processes that we want to foster in our students, as motivation is in fact the fuel of self-regulation. Motivation is best viewed as an ongoing cycle, as seen in Figure 4.1, that can be catalyzed into action by self-monitoring. If we create the conditions for students to set clear goals and benchmark their progress, then they will have greater self-awareness when progress is made. Making progress feels good—it breeds a sense of self-satisfaction that further feeds our motivation. These feelings of self-satisfaction contribute to how we situate ourselves within a field and can lead to a greater sense of personal efficacy—a belief that we can set goals, accomplish tasks, and grow within the field. This sense of efficacy increases the likelihood that we will reengage in new goal setting and self-monitoring and thus further perpetuate the motivation cycle. In short, when students are motivated by their own efficacy, rigorous tasks feel possible and worthwhile.

This cycle of motivation is a framework that we should keep front of mind when we are designing systems in our classroom and courses. If these subprocesses and their explicit linkage are what determine motivation in our students, then we should ensure that the structures of our classes/courses aim to reinforce each. This cycle suggests, for instance, that practice is something that should be approached iteratively, whereby students can act on their growing senses of personal efficacy. This means that whatever vehicles we use for students to engage in practice—be it quizzes, homework, or otherwise—should be deployed in such a way that students are able to revisit, refine, and progress their work. Further, we should recognize that this motivation cycle urges us to promote self-comparison within our students. The *where was I* versus *where am I* comparison should always yield positive results for each learner on a given topic over time. The fact that learning is happening is something that should be both celebrated and *owned* by each student, as, again, this sense of self-satisfaction fuels personal efficacy. Simply put, knowing how motivation manifests in our students can be a powerful means of perpetuating it in our classrooms and fostering the conditions for self-regulation.

CHAPTER 4 SUCCESS CRITERIA CALLOUT:

☐ I can delineate self-regulation into its three phases and describe each.

Zimmerman (2002) also breaks down the structure and function of self-regulatory processes into three phases: the forethought phase, the performance phase, and the self-reflection phase. The forethought phase includes processes and beliefs that occur before a learning effort, including task analysis—where students engage in goal setting and strategic planning—and self-motivation beliefs—such as self-efficacy, outcome expectations, intrinsic interest and value, and learning goal orientation. The performance phase includes occurrences during a learning effort, such as self-control—which includes imagery, self-instruction, attention focusing, and task strategies—and self-observation—including self-recording and self-experimentation. The self-reflection phase involves self-judgment—which includes self-evaluation and causal attribution—in addition to self-reaction—which includes a sense of self-satisfaction or change in affect, as well as adaptive or defensive responses (Zimmerman, 2002). These three phases and their subprocesses are illustrated in Figure 4.2.

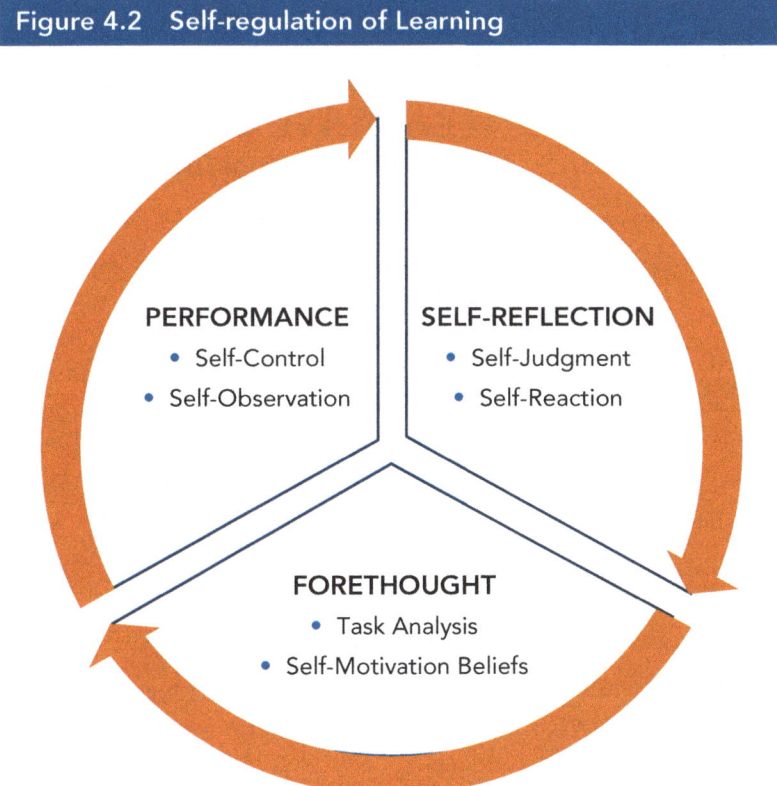

Figure 4.2   Self-regulation of Learning

qrs.ly/vkfo929

Scan the QR code for a more detailed figure of self-regulated learning.

To read a QR code, you must have a smartphone or tablet with a camera. We recommend that you download a QR code reader app that is made specifically for your phone or tablet brand.

Source: Adapted from Zimmerman (2002).

The research on self-regulation of learning offers some recommendations in order to target each of these phases and subprocesses for development with case-specific scaffolding. First, it suggests that when appropriate, students should be provided choice with academic tasks, methods of completing assignments, and peer groupings. It also encourages teachers to establish learning goals (i.e., learning intentions and success criteria) and teach study strategies. Students should also be provided with opportunities to self-evaluate and self-assess their competencies. All the while, teachers should be assessing students' beliefs and efficacy in order to aid in motivation

(Zimmerman, 2002). Importantly, Zimmerman (2002) notes that "[e]ach self-regulatory process or belief . . . can be learned from instruction and modeling." In the next section, we will work to manifest many of these recommendations with strategies suitable for the mathematics classroom. Four instructional strategies that promote self-regulation are goal setting with accountability, using success criteria, giving formative quizzes, and doing practice math assessments.

CHAPTER 4 SUCCESS CRITERIA CALLOUT:

☐ I can build opportunities into my course for students to explicitly engage in self-regulation of learning.

☐ I can create opportunities for students to engage in self-assessment.

Quick Write—Check-in—Exit Ticket: Goal Setting With Accountability

The three part *Quick Write—Check-in—Exit Ticket* strategy is intended to walk students through all three phases of self-regulation in sequence by making each explicit and palpable. When using this strategy, the teacher begins the day with a quick writing activity where students respond to a brief prompt about goal setting. Potential examples include the following:

- How will you contribute to class discussions today?

- What are your goals for working with your group today?

- How do you want to respond today when the math gets challenging? What strategy do you want to try when you don't know what to do?

- What standard are you going to hold yourself to when explaining your thinking verbally or in writing? What will your complete responses look like or sound like?

This initial quick write can occur as a warm-up activity immediately as students walk in the door or after the day's learning intentions have been discussed, based on the structure and goals of the lesson. This initial activity serves to engage students in the forethought phase of self-regulation. Then, once the lesson is at its approximate midpoint (or a nearby natural pausing point), students are redirected to their initial goals they set for themselves and prompted to "check in." Check-ins can occur through another quick write or via a simple ranking system where students can determine the degree to which they are making progress toward the goals they set for themselves. This highlights the performance phase of self-regulation for students and prompts them to self-observe. Finally, as the day's lesson concludes, I suggest closing the loop for students by providing them an exit ticket focused on self-reflection—the third phase of self-regulation. A simple prompt, such as *Did you meet your goals? Why and how, or, why not?*, can spur students to self-evaluate and attribute causation.

Some teachers might choose to build this strategy into their instruction regularly and have students keep journals recording daily goals and progress integrated into their other math notes and work. The benefit of this organization method is that students can revisit their goals and progress over time, which could help perpetuate motivation as defined earlier in this chapter. Others might want to keep their students' responses loose so that they may be collected and responded to the next day. In this second approach, the benefit is similar to that of any other exit ticket: Teachers can sift through student data for trends and patterns that might inform their instruction. Regardless of how, specifically, this strategy is rolled out, the value comes from habitualizing students' explicit awareness of and reckoning with the forethought, performance, and self-reflection phases of self-regulation.

Younger students can engage with this strategy using a template such as shown in Table 4.1.

Table 4.1   Quick Write-Check In-Exit Ticket Sample Template

Today my math goal is:

How am I doing so far?

☺ ☺ ☹ 😐

How did I finish?

☺ ☺ ☹ 😐

Source: Smiley face icons courtesy of iStock.com/Makrushka

 Available for download at **https://companion.corwin.com/courses/whosemathisit**

Success Criteria: Setting, Signaling, and Closure

I stressed the importance of teacher clarity at the onset of this book—we have to know what we want for our students if we are to operate with intention. The goals that we set for our students, however, should not be kept secret for only us to behold. Much of the value of teacher clarity is realized when it is leveraged toward student clarity and empowerment. When we can engage students in iteratively monitoring their learning, they are, by definition, self-regulating. Success criteria are an excellent tool to aid in this process. Success criteria communicate to students what it means to successfully learn a segment of content. In mathematics, this could mean demonstrating knowledge of a concept, mastery of a skill, or the ability to apply mathematical knowledge to new and novel situations. Many teachers choose to communicate success criteria in the form of student-facing *I can* statements, as is shown in this text. Table 4.2 and Table 4.3 demonstrate a selection of related success criteria—written as *I can* statements—from each of the aforementioned mathematical categories.

Simply determining and sharing success criteria, however, does not ensure they are mutually understood. As researchers Nicol and Macfarlane-Dick (2006) put it, "Students can only achieve learning goals if they understand those goals, assume some ownership of them, and can assess progress." And while, these researchers note, feedback can be a reactive means of clarifying success criteria to students, more proactive measures can be taken at the onset of an assignment or task. In addition to providing students with verbal descriptions and written statements of success criteria, research suggests providing complementary strategies for clarifying and interacting with success criteria (Nicol & Macfarlane-Dick, 2006). Exemplars of completed mathematical work serve as one such strategy in that students may actively compare their working progress to a static example of successful completion. Teachers can also carefully construct their written success criteria to include performance level definitions, much like a rubric. And rather than simply describing each of the criteria to students, teachers can invite discussion and reflection about them before students begin

Table 4.2   Examples of Related Success Criteria Involving Two-Digit Addition Without Regrouping (Elementary)

CONCEPTUAL UNDERSTANDING	PROCEDURAL SKILLS	APPLICATION
☐ I can explain the value of each digit in a two-digit addition problem. ☐ I can use base-ten blocks to represent each addend in a two-digit addition problem and show how to use them to find the total.	☐ I can add 2 two-digit numbers using the expanded form model. ☐ I can add 2 two-digit numbers using a number line model. ☐ I can add 2 two-digit numbers using the standard algorithm.	☐ I can mathematize situations involving addition by identifying important quantities. ☐ I can solve real-life problems involving two-digit addition.

Table 4.3   Examples of Related Success Criteria Involving Linear Velocity and Angular Velocity (Secondary)

CONCEPTUAL UNDERSTANDING	PROCEDURAL SKILLS	APPLICATION
☐ I can compare and contrast linear velocity and angular velocity and explain each. ☐ I can demonstrate linear and angular velocity with a simple spinning model.	☐ I can calculate the linear velocity of a point moving with uniform circular motion. ☐ I can calculate the angular velocity of a point moving with uniform circular motion. ☐ I can convert linear velocity to angular velocity and vice versa.	☐ I can mathematize situations involving circular motion by identifying important features and quantities (radius, linear velocity, angular velocity, linear measurements, angular measurements, time measurements). ☐ I can solve real-life problems involving linear and angular velocity.

their work. Research also suggests facilitating peer-assessment against success criteria (discussed in an earlier chapter), which could spark metacognition in students as they assume the role of the evaluator with a peer's mathematical work. The final suggested strategy involves co-constructing success criteria for a piece of work or assignment along with students in a negotiating process when appropriate. Each of these strategies is intended to enhance the clarity of success criteria and thus empower students to more effectively use them as self-regulated learners.

Positioning students in an ownership role with success criteria helps enhance the quality of their engagement with the *forethought phase* of self-regulation. The value of success criteria toward self-regulation does not end there, however. At key points throughout a lesson, task, or activity, teachers can signal to students which success criteria they should be advancing toward in order to promote self-regulation in the *performance phase*, as well. Some teachers choose to do this by revisiting success criteria in their lesson slides or by regularly referring to lesson cover pages where the day's criteria are housed, as seen in Figure 4.3. Other teachers choose to signal specific success criteria to their students by including callouts that segment a lesson's content for students. These callouts can occur before the relevant content pertaining to given criteria (as I have modeled throughout this text) in order to prime the learner for upcoming information. Alternatively, success criteria callouts can occur after relevant content in order to signal intended completion of a learning segment. Callouts of this second type serve as self-assessment mile markers for students as they progress through a lesson. Figure 4.4 demonstrates this latter form of signaling from our colleague Maggie Fallon's Integrated Mathematics 1 course. The sample in Figure 4.5 comes from a portion of printed classwork that drives the day's lesson with whole-group, small-group collaborative, and individual components built in. Ms. Fallon often refers to these success criteria callouts during lessons and reminds students to check in with their learning: "Okay everyone, so we just finished a problem in which we found the slope and we found the y-intercept. At this point, if you feel ready, go ahead and check that box signifying that you have met that success criterion! If you don't feel ready, that's okay, too. We will have many more opportunities to practice and make sure we feel confident enough to check that box." Signaling of this type can provide an organizational structure for students that might not have been self-constructed otherwise. Additionally, structures such as these in our classes establish patterns that aid students in further internalizing the self-regulation process.

Figure 4.3 Sample Secondary Student-facing Lesson Cover Page (Includes Learning Intentions and Success Criteria to Be Revisited Throughout a Lesson)

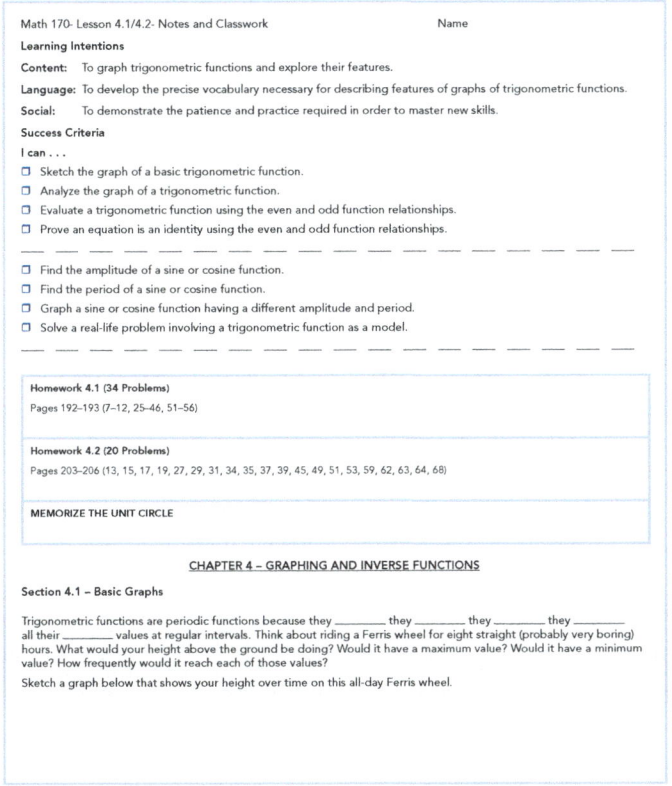

Math 170- Lesson 4.1/4.2- Notes and Classwork   Name

Learning Intentions

Content: To graph trigonometric functions and explore their features.

Language: To develop the precise vocabulary necessary for describing features of graphs of trigonometric functions.

Social: To demonstrate the patience and practice required in order to master new skills.

Success Criteria

I can . . .

☐ Sketch the graph of a basic trigonometric function.

☐ Analyze the graph of a trigonometric function.

☐ Evaluate a trigonometric function using the even and odd function relationships.

☐ Prove an equation is an identity using the even and odd function relationships.

☐ Find the amplitude of a sine or cosine function.

☐ Find the period of a sine or cosine function.

☐ Graph a sine or cosine function having a different amplitude and period.

☐ Solve a real-life problem involving a trigonometric function as a model.

Homework 4.1 (34 Problems)

Pages 192–193 (7–12, 25–46, 51–56)

Homework 4.2 (20 Problems)

Pages 203–206 (13, 15, 17, 19, 27, 29, 31, 34, 35, 37, 39, 45, 49, 51, 53, 59, 62, 63, 64, 68)

MEMORIZE THE UNIT CIRCLE

CHAPTER 4 – GRAPHING AND INVERSE FUNCTIONS

Section 4.1 – Basic Graphs

Trigonometric functions are periodic functions because they _____ they _____ they _____ they _____ all their _____ values at regular intervals. Think about riding a Ferris wheel for eight straight (probably very boring) hours. What would your height above the ground be doing? Would it have a maximum value? Would it have a minimum value? How frequently would it reach each of those values?

Sketch a graph below that shows your height over time on this all-day Ferris wheel.

online resources ⬆ Available for download at **https://companion.corwin.com/courses/ whosemathisit**

Similar to how success criteria can be used to help students self-assess their progress during the performance phase of self-regulation, they can also be operationalized during a lesson's closure. This can, of course, be as simple as referring students back to their daily success criteria and prompting a self-check. This process can be helped along by providing students with referential experiences (such as example problems aligned to each success criteria, short question prompts, etc.) to aid in their self-assessment. The goal of lesson closure of this type is to provide students with a reflective space that explicitly promotes the *self-reflection phase* of self-regulated learning, thus giving them the opportunity to engage in the subprocesses of self-judgment (*Did I learn what I needed to? Why is this?*) and self-reaction (*How do I feel about this?*). This

Figure 4.4   Sample Secondary Classwork With Success Criteria

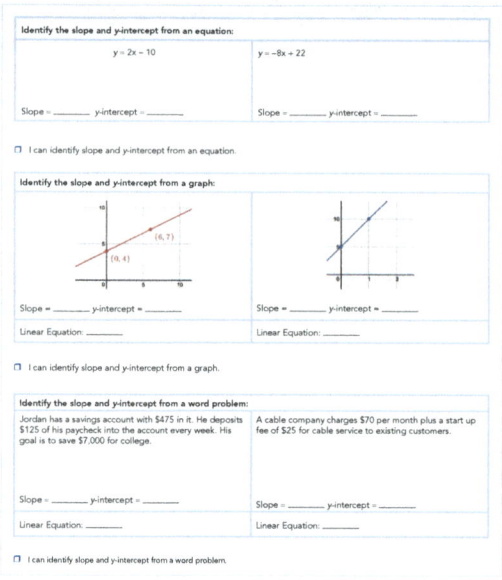

Source: Used with permission by Maggie Fallon.

Figure 4.5   Sample Elementary Classwork With Success Criteria

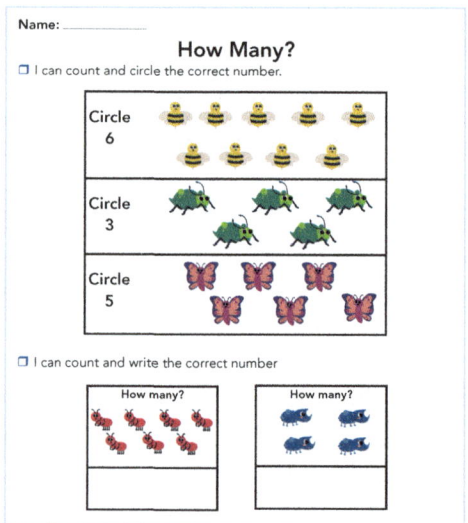

Source: Salomon, M., White Colorful Bugs Worksheet, Canva.com, accessed November 29, 2023, https://www.canva.com/.

online resources   Figures 4.4 and 4.5 are both available for download at **https://companion.corwin.com/courses/whosemathisit**

degree of self-comparison—when focused on learning and individual progress—is quite healthy (and necessary) for inducing ownership in students as they become self-regulated learners.

Formative Quizzes

Why do we give quizzes (if we give them at all)? Quizzes, in the most distilled

CHAPTER 4 SUCCESS CRITERIA CALLOUT:

☐ I can articulate the difference between formative and summative assessments in order to ensure intentional and impactful implementation.

sense, are intended to be formative assessments. Formative assessment, by definition, is intended to *inform* stakeholders how learning is *progressing*. Oftentimes in practice, the space of formative assessment can be usurped by smaller segments of premature summative assessment—the assessment equivalent to a wolf in sheep's clothing. What I mean is this: The distinguishing characteristic between summative and formative assessment is how we choose to use the resulting data. If our "quizzes" are simply scored and entered into the gradebook, then they are in fact summative rather than formative in nature. In order for our quizzes to truly be formative assessments, they should give students and us information so that additional learning can be targeted and ultimately occur. Here are some strategies to that end.

Align quiz items to success criteria. This practice acts as yet another opportunity to transfer our teacher clarity into student clarity. Students only benefit from success criteria if they use them, and quizzes that use success criteria as an organizing framework promote exactly that type of interaction. Further, when we highlight which success criteria are addressed by each quiz item, we create a more continuous and consistent experience for our students by linking otherwise isolated items to larger learning goals. In this sense, if a student struggles with a given quiz item, there exists automatic linkage—via the cited success criterion—to other course materials and experiences for them to review, thus empowering them to take ownership where their learning is unfinished. Figures 4.6 and 4.7 show sample quizzes with this design for secondary and elementary grade levels. Referring to Figure 4.6, if a student struggles with Item 3, for instance, then they are more aware about how that holds greater implications to their ability to *find the exact value of a trigonometric function for a special angle*. With well-designed instructional materials, this language can guide the students where to review. Additionally, well-written success criteria will often yield a plethora of resources when searched online, thus offering students more autonomy over how to finish their learning.

Figure 4.6   Sample Success Criteria Aligned Quiz (Secondary)

Math 170 – Quiz 3 – Section 2.1, 2.2, 2.3p1   Name

☐   I can find the value of a trigonometric function for an angle in a right triangle.

1.   Triangle ABC is a right triangle with C = 90°. If a = 16 and c = 20, what is sin A ?

☐   I can use the Cofunction Theorem to find the value of a trigonometric function.

2.   According to the Cofunction Theorem, which value is equal to sin 35°?

   a. csc 25°   b. csc 55°   c. cos 35°   d. cos 55°

☐   I can find the exact value of a trigonometric function for a special angle.

3.   Which of the following statements is false?

   a. sin 30° = $\frac{\sqrt{3}}{2}$ b. sin 0° = 0

   c. cos 45° = $\frac{\sqrt{2}}{2}$ d. cos 90° = 0

☐   I can use exact values to simplify an expression involving trigonometric functions.

4.   Use exact values to simplify 4 cos² 30° + 2 sin 30°.

☐   I can add and subtract angles expressed in degrees and minutes.
5.   Subtract (67° 22′) – (34° 30′).

☐   I can convert angles from degrees and minutes to decimal degrees or vice-versa.
6.   Convert 76° 36′ to decimal degrees.

☐   I can use a calculator to approximate the value of a trigonometric function.
7.   Use a calculator to approximate sec 31.7°.

☐   I can use a calculator to approximate an acute angle given the value of a trigonometric function.
8.   If cot θ = x for some value x, what would you type into the calculator to solve for θ?

---

online resources 🖱   A full version of this quiz is available for download at **https:// companion.corwin.com/courses/whosemathisit**

Implement quizzes in an iterative fashion whereby students are expected to engage in quiz corrections. I suggest a first round of scoring communicated to students through the "if this were a test" lens. This round can provide formative data to teachers that can inform everything from the need for whole-class reteaching experiences, to small-group instruction, or to individual interventions. Further, this first round of scoring should produce feedback for students that they then have an opportunity to digest and respond to. Quiz corrections can be structured in a way that allows students to reflect on prior thinking, compare it to adjustments they have made since, and discuss why that is of value (and, of course,

---

Figure 4.7   Sample Success Criteria Aligned Quiz (Elementary)

Name: _____

Unit 3 Quiz

Directions: Write the missing number to make the equation true.

[ ] I can find a quantity to make an equation true.

1. $7 + \underline{} = 27$ 3. $18 = 9 + \underline{}$

2. $4 + \underline{} = 54$ 4. $15 = \underline{} + 6$

Directions: Circle the missing symbols to make the equation true.

[ ] I can find a symbol to make an equation true.

5. $23 \underline{} 10 = 10 \underline{} 23$ A. +
                                                                       B. −

6. $15 \underline{} 8 = 10 \underline{} 3$ A. +
                                                                       B. −

7. $6 \underline{} 4 = 4 \underline{} 6$ A. +
                                                                       B. −

---

online resources Available for download at **https://companion.corwin.com/courses/whosemathisit**

recoup missing points if these are graded assignments). The emphasis, I would argue, should be placed on what *changed* with a student's thinking, that is, the *learning* that occurred. Table 4.4 collects the modules from the secondary math courses that I teach that explain the process of quiz revisions to students. This student-facing document is reviewed once at the onset of my courses and then again after the first quiz is scored with feedback (see Figure 4.8).

At the elementary level, you may not choose to have students engage in the same type of quiz revision. Or you may modify how students make revisions to be more appropriate for your students. Regardless, it's important for students to understand the purpose of the different ways to practice/learn new content in order for them to build ownership and purpose in the math classroom. Figure 4.9 shows a sample anchor chart used to explain the purposes of classroom, homework, and quizzes.

Table 4.4   Secondary Quiz Procedures and Corrections

WHAT IS THE POINT OF QUIZZES?

Great question! Without the answer, it will be hard to find value in this experience.

The purpose of **classwork** is to get oriented with new content.

The purpose of **homework** is to build fluency through practice.

The purpose of **quizzes** is to determine your mastery of content *before* the high stakes exams, so that you have the time and information you need in order to respond. Quizzes tell us what we have learned and, more importantly, what we haven't learned. We are already ready to test what we have learned, but we have to target and address what we haven't learned before we can demonstrate mastery.

HOW SHOULD I TAKE QUIZZES?

Even if quizzes are not proctored, you should pretend and take them as if they were a test. This is **your** opportunity to learn what you know and what you don't. You will be initially marked down for incorrect responses, but do not perseverate on the grade—you will have a chance to earn back points!

HOW SHOULD I TAKE QUIZZES? (continued)

Set yourself some rules similar to the test:

1. No notes or resources

2. Only use calculators as directed/allowed (exact answers are almost always preferred to calculator-estimated decimals)

3. Set a timer for 30 minutes. No quiz should take longer than a half hour if you are properly prepared.

Abiding by these self-imposed rules will help you understand whether or not you are ready to demonstrate mastery on each topic.

WHAT DO I DO AFTER A QUIZ?

Make sure you submit your quiz as a single PDF, with everything oriented right-side up. This will make grading and feedback faster (and thus better) for you.

Once I have your quiz, I will provide feedback on your actual document and score your work. This is important information for you to respond to during your final step—quiz corrections (if necessary).

Once you have your graded quiz back, you may correct any problems that were marked wrong and resubmit them for credit.

Process for Quiz Corrections:

1. <u>Completely redo</u> any problem marked wrong on new paper, not on the original quiz.

2. Add annotations next to each problem explaining <u>what you did incorrectly the first time</u> **AND** <u>what you did differently this time.</u> This reflective piece is really important for retention of previously unlearned content.

3. Resubmit the corrections through the original quiz submission link.

EXAMPLE OF QUIZ CORRECTIONS

The following is an example of one of your peer's quiz corrections. Please use it as a reference for what is expected.

Figure 4.8   Student Corrections Example

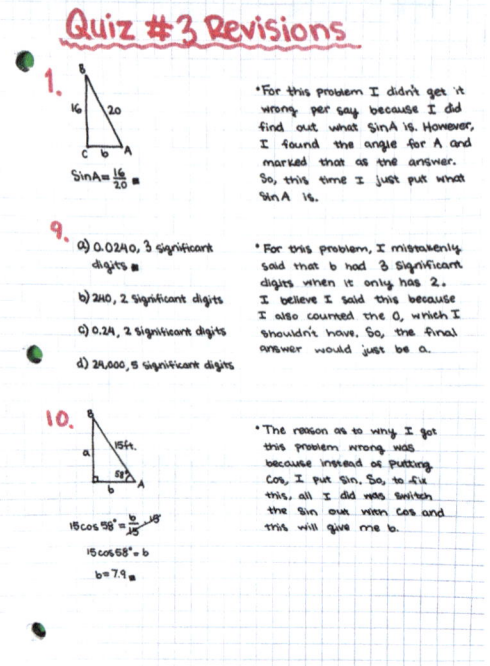

Figure 4.9   Sample Elementary Anchor Chart

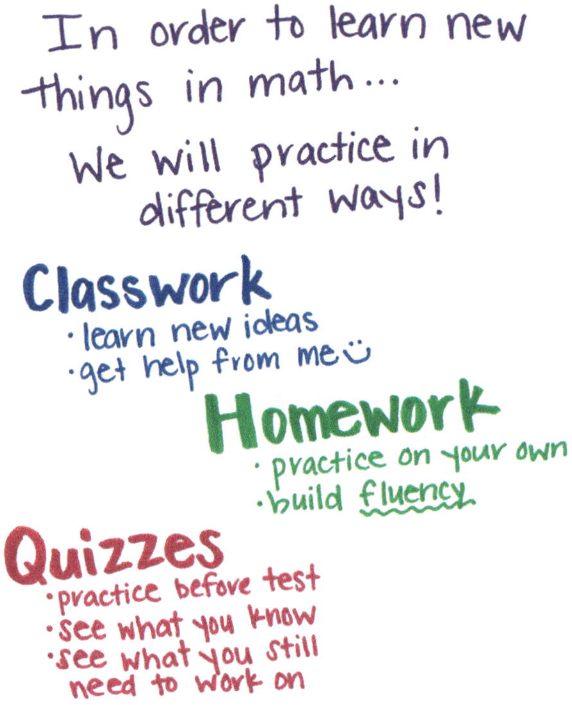

In order to learn new things in math...

We will practice in different ways!

Classwork
- learn new ideas
- get help from me ☺

Homework
- practice on your own
- build fluency

Quizzes
- practice before test
- see what you know
- see what you still need to work on

Practice Math Assessments

Practice math assessments are perhaps one of the most effective ways of engaging students in self-regulation of learning because of their direct nature and the blatant data they provide. Students can use practice tests in faux testing situations to *feel* the impact of a real test before it happens and while there is still time to respond. In order to help students self-generate the most actionable data, consider launching practice tests in intentional phases.

1. Provide students with some independent time to engage with the practice test as if it was the actual assessment. This likely means no supplemental resources. During this first phase, we should instruct students to pay special attention to what they can do with relative ease and note that they are ready to test on that content.

2. Prompt students to continue working independently but with use of their class notes and other resources. In this phase, ask students to identify what content they needed to refer to their notes in order to completely access. This is the content where they need to build some additional fluency and familiarity through practice.

3. Have students work together in order to complete any unanswered questions and share their thinking with one another. Any questions that required this level of support will require extra learning and the most significant amount of attention before the exam.

4. Have students synthesize their experience (rate themselves, rank themselves, back it up with *why*) to create a customized study plan, and help them enact that plan. Study plans and study skills will be discussed in more detail in the next section.

Before moving further, I would be remiss not to share this cautionary note on designing practice tests. Some content should be completely demystified for students. If we are assessing procedural skills or conceptual understanding, for instance, we should not be vague with our students. In fact, I would argue, the more targeted and clearer we are, the better. If we are assessing problem-solving, then the problem should be quite apparent to students. Problem identification, however, is a different story. The process of mathematization, which is the application of mathematics to new and novel situations, requires exactly that—new and novel situations. If we routinize a specific type of word problem, for instance, we run the risk of assessing procedures masked as application. Now, this is not to say that we don't ever want to teach students how to deconstruct a context

> Is it truly mathematical ownership if students can only mathematize problems of the exact types and structures that we have shown them?

or sensemaking strategies—we very much do! I am simply stating, however, that we cannot assess sensemaking skills on a problem that students have already completely proceduralized (or, worse yet, that we have proceduralized for them). For, is it truly mathematical ownership if students can only mathematize problems of the exact types and structures that we have shown them?

One of the ways that I have walked this line with students is through using a two-part test structure. The first part of the test is focused on concepts, procedures, and some light application, and is worth 90% of students' grade. The second and shorter section, worth the remaining 10% of students' grade, is a menu of application problems the students may choose from to demonstrate their mathematization skills. My practice tests are in lockstep with the first portion of the actual exam—no secrets there! In order for students to prepare for the second section, however, I refer them to their classwork experiences and homework exercises. I am not trying to surprise them on this last portion of the exam. I am, however, trying to protect its novelty so that I can authentically assess each student's ability to engage in problem identification and application.

While formative quizzes, practice exams, and the like are all excellent tools to help students recognize *what* they need to study, the question of *how* they should study remains. In the next section we will explore a number of goal-specific study skills for mathematics that can be explicitly taught to students. When students are empowered to become their own teachers through effective studying, their ownership of academic content becomes boundless.

Study Skills for Mathematics

Study skills are strategies that students use in order to learn and retain knowledge. These techniques help them both prepare for and complete schoolwork and tests. Study skills are highly customizable and can include things like test-taking strategies, time management skills, note-taking practices, reading and memorization techniques, amongst a multitude of others. Hattie (2009) identified the impact of study skills on student learning to be well above average and categorized them across three domains: cognitive, metacognitive, and affective.

 **CHAPTER 4 SUCCESS CRITERIA CALLOUT:**

☐ I can identify multiple study skills that are appropriate for mathematics and teach my students to use them effectively.

☐ I can create opportunities for students to engage in self-assessment.

Cognitive study skills involve some sort of studying task, such as notetaking, summarizing, repeated reading, or otherwise, and involve students directly engaging with the content. Cognitive study skills fall entirely in the previously described *performance phase* of self-regulation.

Metacognitive study skills pertain to self-management, such as planning and monitoring, and largely lie in the *forethought* and *performance phases* of self-regulation.

Affective study skills involve motivation, agency, and self-concept and tend to take advantage of the *self-reflection phase* of self-regulation. While each category has its benefits and uses, Hattie (2009) also discovered that the high degree of effectiveness of study skills is only realized when the cognitive study skills are specified to a given content area and course of study.

For our purposes, this means that we need to consider and promote mathematics-specific cognitive study skills for our students. So then, based on our definition of success in mathematics—the goal of promoting individual *ownership* of mathematics in our students—how can we *study* in math? Following is a nonexhaustive list of cognitive study strategies specific to mathematics.

Study Skill: Explain Why—Model Your Thinking to Someone

Identifying gaps in understanding can be challenging for students to engage in individually. Understandably so, as this is the space of formative assessment and is typically catalyzed by the teacher. Teaching students to identify where they have unfinished learning and what questions they might need to ask is an invaluable tool in developing conceptual understanding of mathematics. While discussing the many benefits of collaborative learning early on in my career, I remember Nancy Frey impressing upon me that *the standard of knowing is being able to explain your thinking to someone new*. Indeed, think about the difference in cognitive demand between being able to *do* mathematics versus being able to *teach* mathematics. We have all experienced the increased challenge of teaching a specific concept for the first time—we really have to reckon with and reconcile our own conceptual understanding! This potential for cognitive dissonance can in fact be harnessed as a study tool using the *Explain Why* method.

With this study skill, students challenge themselves to explain a specific math problem from start to finish to someone else. If no one else is available (or willing), they can use a pet, the mirror, or even just speak it out loud. The importance here is that they are verbalizing their thinking in order to manifest potential gaps in understanding that might be internally glossed over. Special emphasis should be placed on what it means to *explain why* rather than *tell how*. Telling how something happened is essentially reciting or reading what is already written (First I added two to both sides, then I divided by six, etc.). Explaining why, however, involves justifying why every step was taken by modeling one's thinking (My goal here is to solve for *x*, which means isolating it on one side of the equation by itself. First, I noticed that two was being subtracted from the term that has *x* in it, so I added two to both sides to get *x* closer to being by itself while also keeping the equation balanced.). When attempting to model our thinking, it quickly becomes abundantly clear where our understanding breaks down. We sometimes might procedurally know *what* to do without necessarily knowing *why*. Students should be coached to recognize this dissonance and note whatever they can't explain for further review. These self-identified gaps can help students decide what they need to ask their teachers, peers, or look up independently in order to finish their learning.

Study Skill: Deliberate Practice

Warning: the following information is not novel or revolutionary, but it is true. The only way to get better at skills is to practice them. Many readers are probably familiar with the 10,000 hour principle, popularized by Gladwell's (2008) *Outliers*, whereby an investment of approximately 10,000 hours of practice is required in order to achieve mastery at a given complex skill. Before panic sets in about the sheer quantity of skills that we must teach our students and severe lack of "hourage" to meet this metric, let's recognize a few things. First, while many of the skills we teach indeed vary in complexity, none are quite as broad in scope as what Gladwell was describing here, such as playing the violin or playing hockey. Arguably, if we were to zoom out to the perspective considered by Gladwell, then the single skill we are charged with teaching is *doing math* or *being a mathematician* (in which case, we have plenty of time!). Second, while the quantity of practice clearly matters, so does the quality as measured by its targeted deliberateness. Perhaps you have experienced an exam situation where you found yourself underprepared despite the fact that you studied, but you studied or practiced the *wrong material* as indicated by your unfamiliarity with the exam. If not, surely you can relate. In order for practice to bear fruit, it must be targeted. This is, yet again, another area where the thoughtful course/unit/lesson alignment to success criteria can be highly beneficial to students. If our assignments, tasks, and tests are aligned to success criteria, and students practice skills aligned to these same success criteria, then their practice will in fact be deliberate, and their studying will manifest results.

This discussion of practice begs us to revisit one of the constructs of mathematical success that we investigated from multiple angles in Chapter 1: procedural fluency. Fluency, sometimes referred to as automaticity, is born from students' repeated exposure to and practice with various concepts and skills. Repeated exposure and practice builds familiarity, and with continued effort, familiarity matures into fluency. Fluency, in a nutshell, is complete ownership of a mathematical skill such that the user can recognize where it applies and how to employ it with minimal effort and without direction. Fluency cannot be purchased, requested, wished, or inherited; it can only be earned. Thus is the purpose of deliberate practice.

Study Skill: Self-Assessment Through Self-Ranking and Benchmarking

Sometimes we tend to oversimplify our assessment of our own mathematical understanding into a binary measure: I know/don't know this concept or I can/cannot do this skill. While this simple measure is more valuable than a complete absence of self-assessment, employing an evidence-based, sliding-scale ranking system instead can prove more actionable for students. Consider the following such scale that aids students in self-assessment.

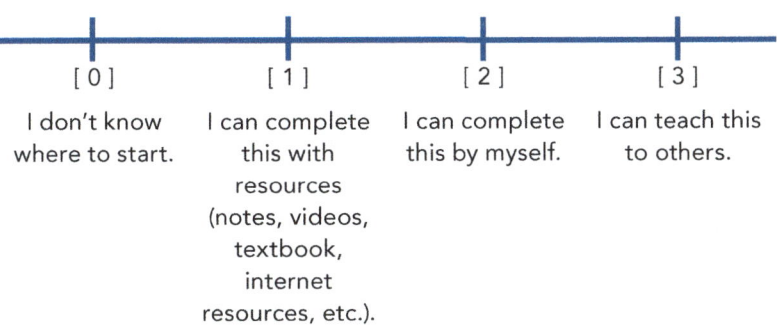

Notice that this scale still provides space for a complete lack of access (Rank 0), which is a signal that the student needs direct intervention and reteaching on that concept or skill. Successful completion of a problem or task, however, is qualified into three categories of varying degrees of independent ownership. This helps students recognize the difference between being able to complete a task through the use of scaffolds (Rank 1) versus through their own fluency (Rank 2). Further, a scale like this pushes students to recognize that previously mentioned difference between what can be completed alone (Rank 2) and what can be taught to others (Rank 3). The latter is evidence of mastery.

A study tool like this scale can be implemented by students much like a workout plan. Identify a starting point, then push to the next level—much like one might add a little weight to a barbell after a successful set with lighter weight, or increase the rate or duration (or both) of their treadmill run. As stated before, if a student doesn't know where to start, they need to get a trainer—to keep with our metaphor—to help get them started. If a student can successfully complete a task with scaffolds and supports, then they should keep practicing similar tasks while trying to refer to notes less and less until they can exercise the skill completely independently. And if a student can successfully complete a task independently, they should try to *explain why* by modeling their thinking to someone else, as described previously. Self-ranking with this type of tool encourages students to benchmark their progress and approach their learning iteratively with an eye on improvement.

Circling back to our ongoing theme regarding teacher clarity and success criteria, with a well-designed course, intended learning outcomes should be of no surprise to students. So, even if we choose not to employ explicit tools such as practice tests, students should still have enough knowledge of which concepts and skills will be assessed by future tests in order to craft their own study plans. Simply providing them with written success criteria, however, does not ensure that they will be able to engage in actionable self-assessment and study-plan design. With training, on the other hand, research has shown that students can accurately assess themselves and take ownership of their learning (Wong, 2014).

Self-assessment should inform studying and test-preparation and should include the following:

1. **Comprehension**—Digesting the meaning of success criteria

2. **Ranking**—Self-ranking the degree to which each success criterion is presently mastered

3. **Action**—Taking specific and intentional action based on the degree of mastery of each success criterion

Figure 4.10 demonstrates how to put these three components of self-assessment into action for students through a study guide template. In this example, students are directed to begin by reading each of a unit's success criteria one by one and identify where they are addressed in their classwork materials. After revisiting each specific success criterion and where it is presented in their own materials, students are told to rank themselves on a scale of 0 to 3 using the sliding scale previously described. Figure 4.11 demonstrates how to modify this for an elementary classroom. Instead of providing students with an extensive guide, you can have students self-rank each success-criterion aligned problem in the margins of a practice test.

Figure 4.10   Secondary Sample Study Guide Planner Example

Unit 3 Study Guide

Your upcoming exam was written based on each of your success criteria for this unit. In order to prepare, you should generate a study guide. Here are some guidelines:

1. Rank yourself from 0-3 with each success criteria using the scale below. Start by reading each success criterion and identifying where it was addressed in the classwork.

[0]	[1]	[2]	[3]
I don't know where to start.	I can complete this with resources (notes, videos, textbook, internet resources, etc.).	I can complete this by myself.	I can teach this to others.

    a. [3] Any success criteria that you ranked a 3 are already **mastered,** and you don't need to do any studying other than maybe a brief skimming before the exam.

    b. [2] If you marked any success criteria a 2, you are **competent** with those skills and might want to consider trying to explain them to others in order to master them.

    c. [1] Success criteria that you marked a 1 are **in progress** and you are **not** ready to test on them. In order to develop further competence, practice similar problems until you can do them without the help of notes or other resources.

    d. [0] If you marked any success criteria with a 0, then you **have not yet learned** them, and you need to engage in new learning. Check in with peers, tutors, teachers, or search key terms for internet support immediately to start making progress.

2. For any success criteria that you marked under a 0 or 1 (and optionally for 2), gather key concepts and examples from your notes to create your study guide. *Remember, study guides should be personalized to include only what you need to study.*

3. Study. Try to move yourself up the scale with repeated practice (try to move 0s to 1s, 1s to 2s and 2s to 3s). Find additional examples in your homework and classwork to try on your own.

Success Criteria

I can . . .

❑ Follow the "chaining" logic of compositions of functions to determine values of composite functions from tables or graphs.

❑ Substitute and simplify various compositions of functions.

❑ Rewrite a single function as a composition of multiple functions.

_____

❑ Use the logic of inverse functions to correlate elements of the domain and range.

❑ Determine whether functions have inverses through the horizontal line test.

❑ Find a function's inverse when given its formula.

❑ Evaluate inverse function values given a graph of a function (read a graph "backward").

❑ Show that two functions are inverses of each other by composition.

❑ Interpret function values and inverse function values in the context of a problem.

_____

❑ Add, subtract, multiply, and divide functions.

_____

❑ Match power functions to their graphs by considering where they are increasing and decreasing, their end behavior, their concavity, and by comparing rates of change.

❑ Find formulas of power functions given two points.

❑ Write formulas from descriptions of directly proportional and inversely proportional relationships and use them to solve problems.

_____

❑ Identify the _leading coefficient_ and _constant coefficient_ in polynomial functions.

❑ Find the formula of a polynomial function given its graph and a few key points.

❑ Determine the number of roots and turning points of polynomial functions.

❑ Determine the end behavior of polynomial functions.

_____

❑ Determine horizontal asymptotes numerically.

❑ Determine horizontal asymptotes graphically.

❑ Determine horizontal asymptotes of rational functions algebraically.

❑ Determine vertical asymptotes numerically.

❑ Determine vertical asymptotes graphically.

❑ Determine vertical asymptotes of rational functions algebraically.

❑ Find x- and y-intercepts of rational function s (if they exist).

Figure 4.11   Secondary Sample Study Guide Planner Example

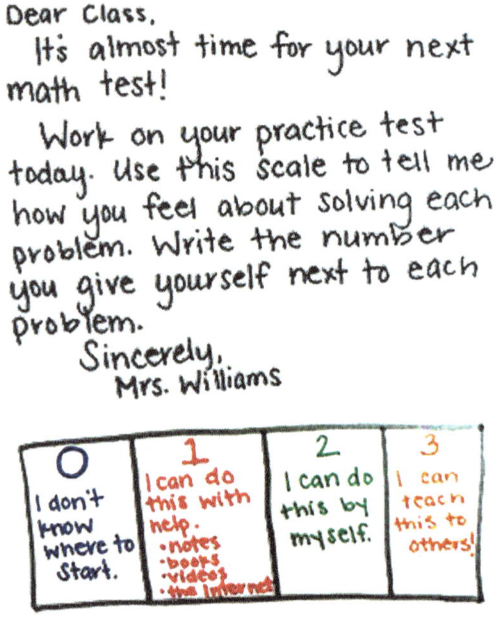

From there, these rankings are translated into implications about present states of learning and suggested courses of action. If students rank themselves a 3 on a given criterion, for instance, this implies mastery and that minimal studying is required. Whereas a ranking of 1 implies learning is in progress, but students are not yet ready to test and should practice similar problems until they are no longer reliant on resources. This sliding mastery scale is not communicated as static but dynamic in that students are encouraged to try to move from one ranking to the next through specific actions and to reengage in self-assessment to determine growth along the way. Tools such as these allow learning to become truly autonomous as students become driven by self-regulation and guided by results. Now students don't just know *what* to study but also *how* they might study. Further, the motivation cycle discussed earlier in the chapter is in full effect with a tool such as this.

Study Skill: Diagnose Mistakes and Errors

Students often have access to some form of feedback with much of the practice we assign them. Many digital programs provide instant feedback, and some even offer step-by-step exemplars. Apps exist that can answer many math problems with the scan of a cell phone camera. Even many print resources have selected answers readily available for students to reference. Further, many (if not most) common textbooks have been turned inside out by online communities with solutions and worked exemplars provided for many exercise sets. Great! Such is our reality. Now how can we train students to make the most of it? We can teach students to use these resources as comparative and diagnostic references. First, instruct students to try a problem or task without use of the resource, as these resources should be used reactively and not dependently. If the answers don't match up, then students can start to compare their work to that of the exemplar, noting where each deviates. If they spot a difference in their work, they should try to determine what they were thinking and why that was the case. There was likely logic there, though possibly misplaced. The idea here is to help students recognize whether they made a *mistake*—a simple slip-of-the-mind and direct consequence of being human, or if there is an *error* in their thinking—a genuine need for new learning due to misunderstanding (Fisher & Frey, 2015). Figure 4.12 compares and contrasts mistakes and errors. If a student recognizes where they made a mistake, this is often a simple fix and the correction serves as

Figure 4.12   Comparing and Contrasting Mistakes and Errors

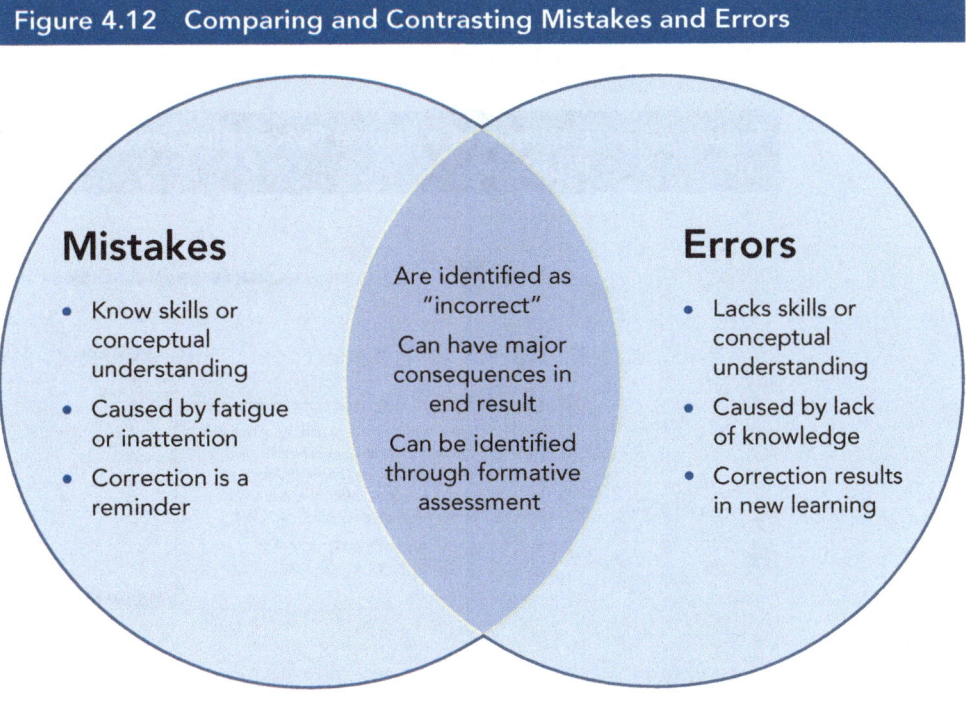

Source: Adapted from Fisher and Frey (2015).

a reminder of what they already know. If, however, a student recognizes an error, their thinking is challenged as they might not see how or why their work is incorrect, even when compared to an exemplar. Discovery of potential errors, I would argue, is what students should be instructed to document, note, journal, or otherwise and bring up with their teacher, tutor, or classmates. A shortlist of corrected mistakes and questionable errors is a fantastic product to come away with from a study session.

Study Skill: Learn Anywhere

It is no secret that amazing learning resources exist for all subjects completely for free and are accessible online. As discussed earlier, these offerings do not and cannot replace teachers. The fact that all learning is personal permanently guarantees the necessity of our profession. These resources can, nevertheless, offer a wonderful in-the-moment form of supplemental support to students. We should not assume, however, that students know how to use these resources as a study tool. As students progress into middle school and certainly in high school, virtual learning resources can help move a study session forward and keep it productive if students know how to find what they need. For instance, imagine that you were outside of regular class hours trying to study and you did not have any access to attempt the problem shown in Figure 4.13.

Figure 4.13   Example of a Typical Precalculus Textbook Problem

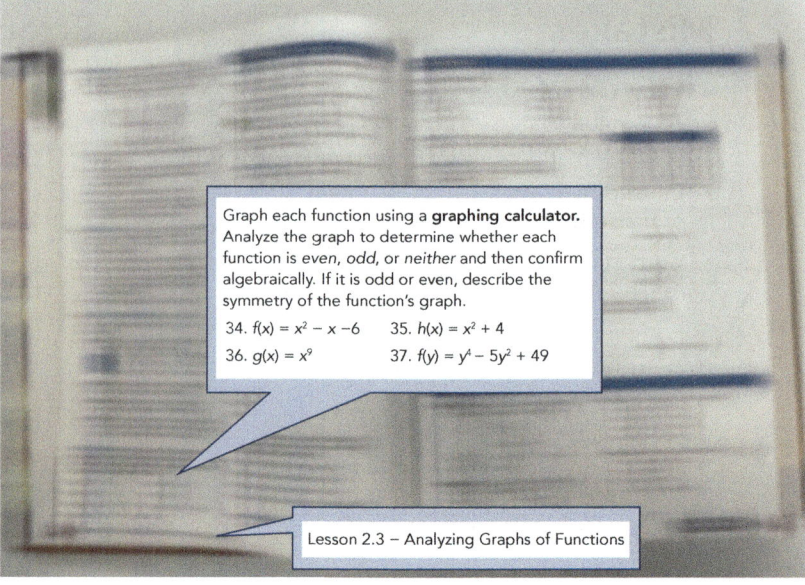

Graph each function using a **graphing calculator.** Analyze the graph to determine whether each function is *even*, *odd*, or *neither* and then confirm algebraically. If it is odd or even, describe the symmetry of the function's graph.

34. $f(x) = x^2 - x - 6$ 35. $h(x) = x^2 + 4$

36. $g(x) = x^9$ 37. $f(y) = y^4 - 5y^2 + 49$

Lesson 2.3 – Analyzing Graphs of Functions

What would you search online in order to get a boost? You would probably start by dissecting the problem language looking for the verbs and nouns: *What am I trying to DO to WHAT?* Perhaps you would look at the section or lesson title for further clues about the content umbrella under which this problem might be frequently categorized. You might even back up to the chapter or other organizing framework to broadly orient your search. Is there key searchable language in success criteria that can lead to additional resources? Further, where would you search? Are there common virtual math resources that you prefer based on experience? I encourage you to model this entire process for your students as a means of empowering them to *learn how to learn anywhere* and thus keep their studying moving forward. Deconstructing a math text (or other context) by investigating its structure is an excellent means of determining what concepts or skills one needs in order to access a given task.

Study Skill: Talk About Math— Storytelling and Concept Mapping

In my master's program, I had a math professor that stressed that every course should have a story. This professor would often return to the "story of the course" to begin lectures—almost like a series recap at the beginning of a new episode of a show (a la, *previously in linear algebra . . .*). The emphasis provided by this professor's format was on connectivity and conceptual understanding. Relevance of contemporary content was enhanced by its clear connection to and place within the course as a whole. In this sense, the course felt like an ever-growing concept map. My study group and I found ourselves emulating this approach in our sensemaking conversations about course content.

Likewise, students can study for a math test by talking about its content and how it is related. Elementary students will likely require more structure and assistance with these conversations, so consider holding these discussions whole-class or in small teacher-guided groups. Prompt students to organize a unit of study or perhaps the entire unit/course into a story in conversation. Instruct them to try to tell the story of the course together, identifying where the story breaks down as an indicator of a lack of conceptual connectivity. Challenge students to identify major concepts in the course or unit—hence exercising their academic and content vocabulary—and discuss how they are related to or build off of one another. All of this serves as self-assessment to help guide them to ask the right questions as the test date gets closer.

Students can also generate concept maps to organize their knowledge of unit/course concepts and their connectivity. A concept map is a graphic organizer that is used to show connections between multiple concepts via lines and accompanying linking phrases. This tool has a number of uses in the classroom as it arguably mirrors the network-like structure of knowledge itself—be it novice or expert. Concept maps provide a window into conceptual understanding that is not necessarily available through other means of communication. They can be an efficient and effective tool for students to organize and articulate their own understanding. Figure 4.14 shows a student-generated concept map using the prompt, *What is trigonometry?* I encourage you to investigate the details of this concept map as a lens into the student's conception of trigonometry. There is evidence that this student's knowledge of trigonometry is developed well beyond the basic level. However, there exist a number of surprising (mis)conceptions here that would not necessarily be made apparent on traditional exam data. Figure 4.15 shows a similar task at the elementary level, where the teacher led a small group to generate a concept map using the prompt, *What is a triangle?* In this approach, the teacher facilitates creation of the concept map by eliciting conversation around the prompt and builds consensus using follow-up questions.

Figure 4.14   Student-generated Concept Map From the Prompt, What Is Trigonometry?

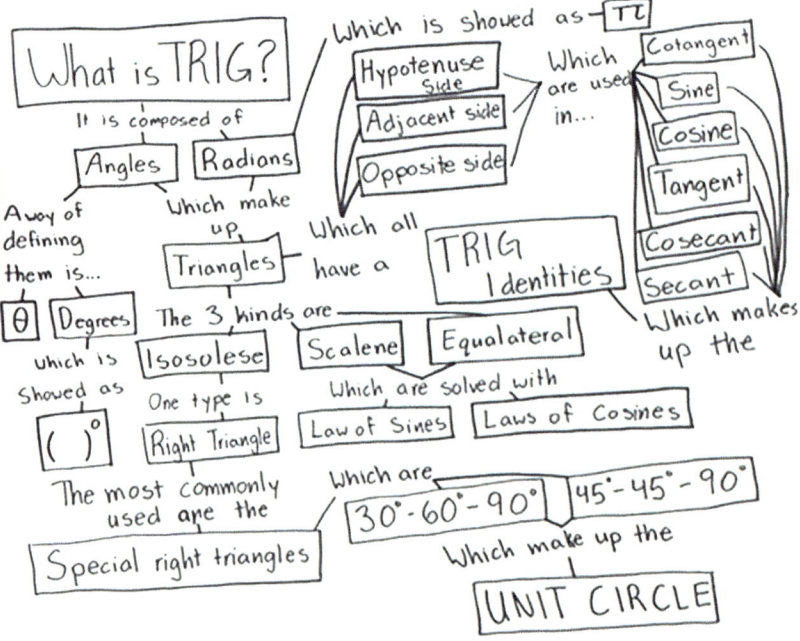

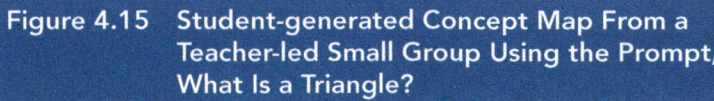

Figure 4.15   Student-generated Concept Map From a Teacher-led Small Group Using the Prompt, What Is a Triangle?

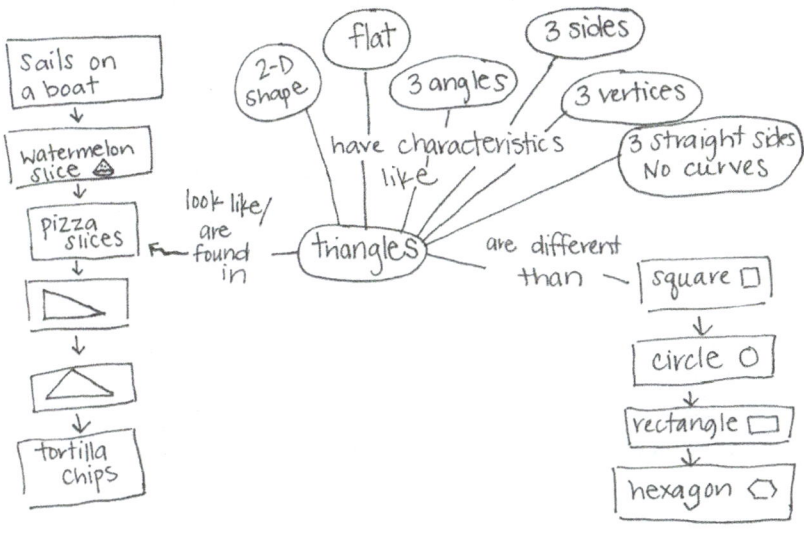

Study Skill: Formulas and Identities— Validation, Understanding, and Recall (Secondary)

Identities and other formulas are a big part of mathematics. These are forever truths that we can leverage in any circumstance without seeking permission. The legitimacy of mathematical identities transcends all circumstances, as their logical purity generates their own authority. In other words, identities always apply! The value of formulas and identities, however, is predicated on the assumption that we know them well enough to bring them to the table on command. With this being the case, we must *own* them. So, how is it that students can study mathematical formulas? One such route is to study identities and formulas in three intentional phases: studying for understanding, studying for recall, and studying for validation.

Studying formulas for understanding places emphasis on their conceptual underpinnings. In practice, studying for understanding is the act of working to derive or generate mathematical formulas from scratch by building off of other logical given truths. This helps to illuminate conceptual connections and reduces the need for memorization. For instance, consider the Pythagorean Identities for trigonometry. With a strong conceptual understanding of the

Pythagorean Theorem and foundational trigonometry, these can all be derived from scratch as follows.

Given the Pythagorean Theorem

$$a^2 + b^2 = c^2$$

and the definitions of the sine and cosine functions

$$\sin\theta = \frac{side\ opposite}{hypotenuse}; \quad \cos\theta = \frac{side\ adjacent}{hypotenuse}$$

we can construct a right triangle and the following relationships for $\sin\theta$ and $\cos\theta$ (see Table 4.5).

Table 4.5   Constructing a Right Triangle

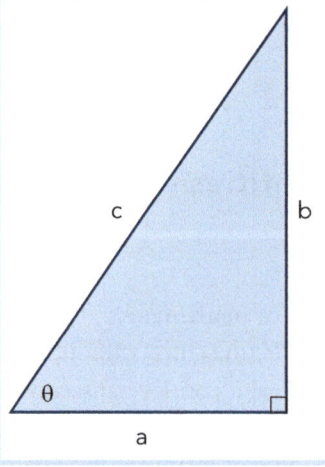

$$\sin\theta = \frac{b}{c} \rightarrow b = c\sin\theta$$

$$\cos\theta = \frac{a}{c} \rightarrow a = c\cos\theta$$

From here we can use the Pythagorean Theorem, substitution, and some algebra to derive the first trigonometric Pythagorean Identity.

$$a^2 + b^2 = c^2$$
$$(c\cos\theta)^2 + (c\sin\theta)^2 = c^2$$
$$c^2\cos^2\theta + c^2\sin^2\theta = c^2$$
$$c^2(\cos^2\theta + \sin^2\theta) = c^2$$
$$\cos^2\theta + \sin^2\theta = 1\ \blacksquare$$

Once we have the first Pythagorean Identity in hand, we can use our knowledge of the reciprocal identities and quotient identities

$$\csc\theta = \frac{1}{\sin\theta}; \quad \sec\theta = \frac{1}{\cos\theta}; \quad \tan\theta = \frac{\sin\theta}{\cos\theta}; \quad \cot\theta = \frac{\cos\theta}{\sin\theta}$$

along with some simple division to generate the others (see Table 4.6).

Table 4.6 Pythagorean Identity Example	
$\cos^2\theta + \sin^2\theta = 1$	$\cos^2\theta + \sin^2\theta = 1$
$\dfrac{\cos^2\theta}{\cos^2\theta} + \dfrac{\sin^2\theta}{\cos^2\theta} = \dfrac{1}{\cos^2\theta}$	$\dfrac{\cos^2\theta}{\sin^2\theta} + \dfrac{\sin^2\theta}{\sin^2\theta} = \dfrac{1}{\sin^2\theta}$
$1 + \tan^2\theta = \sec^2\theta$ ∎	$\cot^2\theta + 1 = \csc^2\theta$ ∎

Based on where students are in their learning, studying formulas or identities for understanding could be a very tall order. In these cases, they should still engage in the process, while noting and ranking their degrees of understanding along the way—much like was discussed in earlier sections. Revisiting proofs or derivations over time can be incredibly fruitful mathematical exercises toward understanding. Sometimes, however, students need to recall and use formulas before they fully comprehend their meaning. This is where studying for recall can be helpful—as long as we don't mistake it for what it is *not*.

Studying for recall is about building students' capacity for memory, not understanding. This distinction is important because it prevents us from miscategorizing and thus falsely replacing conceptual understanding with memorization. Having multiplication tables memorized, for instance, is a fantastic and valuable body of knowledge that will be leveraged countless times throughout a student's mathematical life. Securely holding this body of knowledge, however, does not imply any understanding of the structure or meaning of multiplication. To truly be situated as an *owner* of mathematics, we need both. Strategies toward studying for recall include everything from flash cards to mnemonics and can be applied across grade levels.

Even as students work to develop their understanding and recall of formulas, confidence can be an issue. Studying for validation can boost students forward in these situations. For instance, how can students determine whether their recalled formula actually works? How can they trust their memories, especially if they are still developing their understanding? One study skill that can assist students in validating their less securely-held content is by testing their recalled formulas with sample values. If their formulas hold up to a handful of tests, they are likely correct (though, we should be careful not to confuse this as a formal proof). If, however, students uncover a single counterexample, then they can be confident that their recalled formula is *not* accurate. For instance, consider the Quotient Property of Logarithms. Can you remember it? Perhaps a student recalls that quotients are somehow linked to subtraction, because of the connection to exponents and the fact that whenever two powers are divided their exponents are subtracted. Without reference, however, students might struggle to recall which of the following is a true statement: $\ln\frac{a}{b} = \ln a - \ln b$ or $\frac{\ln a}{\ln b} = \ln a - \ln b$. Testing for validity can help students move forward as shown in Table 4.7.

Table 4.7   Testing for Validity Example

$\ln\frac{a}{b} = \ln a - \ln b$	$\frac{\ln a}{\ln b} = \ln a - \ln b$
Choose $a = 7$ and $b = 2$	Choose $a = 7$ and $b = 2$
$\ln\frac{7}{2} = \ln 7 - \ln 2$	$\frac{\ln 7}{\ln 2} = \ln 7 - \ln 2$
$1.25276\ldots = 1.25276\ldots$	$2.80735\ldots \neq 1.25276\ldots$
Holds true for $a = 7$ and $b = 2$.	Does not hold true for $a = 7$ and $b = 2$.
Choose $a = e$ and $b = 1$	Choose $a = e$ and $b = 1$
$\ln\frac{e}{1} = \ln e - \ln 1$	$\frac{\ln e}{\ln 1} = \ln e - \ln 1$
$1\text{n } e = 1 - 0$	$\frac{1}{0} \neq 1 - 0$
$1 = 1$	
Holds true for $a = e$ and $b = 1$.	Does not hold true for $a = e$ and $b = 1$.

Now, of course we would ultimately prefer students to possess the conceptual understanding that *a logarithm is an exponent* and that its argument is the *output of an exponential function*. For this knowledge would lead to the idea that dividing exponents is not generally the same as subtracting exponents, which is what is falsely argued by the statement: $\dfrac{\ln a}{\ln b} = \ln a - \ln b$. That being said, when we view learning mathematics as an iterative journey, the process of testing and validating formulas and identities along the way becomes an imperative stepping stone toward mastery. Additionally, this process can be completely student-driven and thus further perpetuate ownership.

Problem-Identification and Sense-Making Strategies

At a certain point in everyone's mathematical journey, the first step to any problem or task is . . . *what?* Strategizing how in the world to begin a problem, proof, or other task (or even deciphering what is being asked of us!) becomes a to-be-expected initial undertaking. This final section will touch on two more, very similar, sensemaking strategies that students can be taught in order to tread through these treacherous waters that might otherwise lead to shutdown.

Sometimes students are confronted with tasks or problems that are just wordy enough to confuse them. Perhaps the language used or even the manner in which it is used gives students trouble. In these cases, helping students engage in problem identification can be highly valuable. Students can be taught to dissect tasks and contexts by asking themselves three questions, in no particular order:

- What information do we *have?*

- What information do we *need?*

- What *form* should our answer take?

The first question can prompt students to discover and extract information that is provided by the task at hand, such as known quantities, rates, or problem parameters. The second question helps students identify what necessary information they lack and might have to mathematically generate using the provided information. Answering the final question (which can sometimes be helpful to ask first) allows students to generate a frame for their answer that orients their problem-solving path.

A similar approach to problem identification is the *know-given-find* method. Similar to the three questions, this method first prompts students to identify what they *know* about the task context, such as applicable formulas, relationships, or other potentially useful mathematical knowledge. Reading or rereading a task with this lens in mind can help students take baby steps toward mathematization. Second, students are prompted to determine what information they are *given* by the problem, which, again, speaks to the extractable quantities and other parameters from the text. Third, students are asked to establish what they are actually tasked with *finding*. Much like the three questions, this process serves to orient students toward problem-solving by scaffolding the problem-identification process.

This fourth chapter has offered a number of student-facing strategies for self-regulation, studying, and sensemaking. Each of these strategies is designed to be completely operationalized by individual students so that they may take genuine control over their own learning. No assumption should be made, however, that simply informing them of strategies or asking them to use them will result in ownership. Just like content, process should also be taught, modeled, and scaffolded into use with our students. Our role as teachers is to empower our learners, and teaching them how to *be* independent is part of that process.

CHAPTER 4 REFLECTION QUESTIONS

- How have you engaged students in the components of the motivation cycle? Do you have ways of helping them self-monitor, recognize their progress, celebrate their progress, or attribute their progress to their actions?

- What are some goals you have regarding the use of success criteria in your classes? How are you using them now?

- What types of self-assessment opportunities currently exist for your students? Are there others that you think they would benefit from?

- How do you (or did you) study for mathematics as a student? Is there anything from your experience that you feel would be beneficial for students to learn about?

- What are some other sensemaking strategies that you have leaned on when a problem-solving path was not immediately evident? In other words, *what do you do when you don't know what to do?*

Afterword

Actualizing Student Ownership of Mathematics: *You Can Do It*

I acknowledge the scope and gravity of the task set out for teachers of mathematics by this book. It is my hope that the tools I have laid out throughout each chapter are useful as you work toward achieving this goal across three levels of student experience: classwide, peer-to-peer, and individual. Remember to use these three student-facing mantras as access points for building ownership of mathematics:

Student Mantra 1: Everybody's doing it. Develop, maintain, and leverage a classroom culture that promotes student ownership of mathematics. Create a space where shared thinking is expected, honored, built-up, and built-upon. Reinforce this by continuously shaping social and sociomathematical norms that promote mathematical ownership. Explicitly model and demonstrate the frequent choices made by mathematical thinkers to promote students' comfort with their own decision making. Thoughtfully and inclusively facilitate class discussions using discursive positioning moves to situate all voices as relevant and everyone as a problem-solver.

Student Mantra 2: We're doing it. Reinforce student ownership of mathematics by structuring peer interactions and collaboration. Create strategic, heterogeneous groups that set the stage for collaboration. Prime students for productive group work by considering what makes tasks rigorous and then thoughtfully launch with intention. Emphasize choice and ownership in mathematics by creating opportunities for peers to examine, critique, and reflect on one another's work. Choose and plan for collaborative protocols that suit the social and learning needs of your students.

Student Mantra 3: I'm doing it. Consolidate each student's ownership of mathematics by fostering their self-regulation of learning. Build opportunities into your course or classroom for students to explicitly engage in self-regulation of learning. Create opportunities for students to engage in self-assessment. Implement formative and summative assessments with intention and ensure impact by including students in the assessment process. Teach students how to effectively study for mathematics.

In addition to these student-facing mantras, I urge you to consider five teacher-facing mantras to guide your classroom policies and decision making. Keeping these pillars at the forefront of your thinking will help ensure a mindset focused on developing student ownership of mathematics.

Teacher Mantra 1: When scaffolding, less is more. In the absence of access, no rigor exists. However, while we work to scaffold students toward access, we must also plan for the removal of supplemental supports. Opportunities for independence must be present if we are to promote ownership in our students. Thus, we plan for access, support with intent, and plan for removal. There is an appropriate level of struggle and perseverance that should be everpresent in our classrooms. Students should be doing the lifting. They aren't lifting if the task is too heavy for them, and they aren't lifting if we are doing the work for them. We must pursue that individualized balance for each child. Plan multiple levels of scaffolds and release them *only after* assessing and determining need. Start with the least amount of support and continue to increase assistance only if students continue to demonstrate need. We want to maximize each student's lift. Remember the gradual flow of scaffolds:

$$\textit{Questions} \rightarrow \textit{Prompts} \rightarrow \textit{Cues} \rightarrow \textit{Reteaching}$$

Teacher Mantra 2: Hold the bar. Grade-level content. Grade-level content. Grade-level content. Students should be taught grade-level appropriate mathematics. Prerequisite skills should be assessed and addressed throughout pursuit of grade-level content. As teachers, we can scaffold and support students up to grade level rather than teaching down to gaps in background knowledge. One-off remedial measures such as teaching a "Unit 0" full of prerequisite skills are tempting but rarely successful. These approaches tend to eliminate the relevance in mathematics for students and provide a drastic overemphasis on rote procedures. Instead, think about filling gaps as providing near-context for students to pull from during upcoming instruction on grade-level standards and topics (and include those explicit connections in your planning!).

Teacher Mantra 3: Celebrate and OWN success. Get excited about success! Whether students have an ah-ha! moment, make a discovery, figure something out, or perform well on an assessment, we should celebrate their growth. Call out success as it happens, get excited about it, and help students take ownership of their efforts. Explicitly illustrate ownership to them through cause and effect relationships. (e.g., *You have been working so hard on your adding fractions skills and now you just passed your skills test! How does that feel? You must be proud of your work!*) Sometimes, students will want to attribute their success to us. Accept compliments with gratitude but defer ownership of their success—as that belongs to them. (e.g., Student: *I did well because you're a great teacher.* Teacher: *Thank you, that is kind of you to say. But my job is to empower you and YOU did this.*)

Teacher Mantra 4: Know your content. No matter the grade level(s), dig into the *what* and *why* of what you need to teach. Push yourself to develop a conceptual understanding of ALL the math you are responsible for teaching. Use the strategies in this book to support your own learning when necessary. Reach out to specialists, collaborate with colleagues, and increase your mathematical understanding. When we don't know the content, we miss opportunities to elevate students' thinking. Further, as Russell (1945) stated, "Mathematics, when rightly viewed, possesses not only truth, but supreme beauty . . ." We are math teachers. We should pursue this beauty and share the joy we find in mathematics with our students. Teacher dynamism and excitement about their content can go a long way in promoting interest and relevance in students. We owe it to our students to show them just how awesome math is as a field!

> When we don't know the content, we miss opportunities to elevate students' thinking.

Teacher Mantra 5: Listen to students, and genuinely explore their thinking. This can feel risky and is never preplanned. But, give your kids the space to think out loud, share their ideas, and learn in public. The more you practice this, and the stronger command you have over your content, the more fruitful (and fun!) these experiences will become for all. Mathematics is the universal language and literally surrounds us whether we recognize it or not. Coupling this with the fact that all learning is personal makes the value of allowing students to explore their thinking become blatantly apparent. Students bring unique, math-packed experiences to our classrooms. Our role is to help them recognize and unpack those experiences, thus leveraging the currencies they arrive with. A culture of shared thinking is a space where all can thrive.

At its core, individual mathematical ownership is about broadening student independence and sharpening self-advocacy. Students can be taught to become less reliant on their teachers and ultimately categorize them as just another learning tool. When this happens, when we become less needed, our mission is accomplished. When needs and wants become the limiting factors of one's mathematical journey, we have succeeded as teachers—for, at that point, our students truly possess ownership of mathematics.

References

American Psychological Association, Coalition for Psychology in Schools and Education. (2015). *Top 20 principles from psychology for preK-12 teaching and learning.* http://www.apa.org/ed/schools/cpse/top-twenty-principles.pdf

Bloom, B. S., Engelhart, M. D., Furst, E. J., Hill, W., & Krathwohl, D. (1956). *Taxonomy of educational objectives. Volume I: The cognitive domain.* McKay.

Bloom, B. S., Masia, B. B., & Krathwohl, D. R. (1964). *Taxonomy of educational objectives volume II: The affective domain.* McKay.

Bolkan, S. (2017). Instructor clarity, generative processes, and mastery goals: Examining the effects of signaling on student learning. *Communication Education, 66*(4), 385–401. https://doi.org.libproxy.sdsu.edu/10.1080/03634523.2017.1313438

Bunce, G. (2003). *Educational implications of Vygotsky's zone of proximal development on collaborative work in the classroom.* https://www.academia.edu/3753166/vygotsky_and_the_classroom

Cobb, P., & Yackel, E. (1996). Constructivist, emergent, and sociocultural perspectives in the context of developmental research. *Educational Psychologist, 31*(3/4), 175–190.

Common Core State Standards Initiative (CCSSI). (2020). *Key shifts in mathematics.* https://www.thecorestandards.org/other-resources/key-shifts-in-mathematics/

Fendick, F. (1990). *The correlation between teacher clarity of communication and student achievement gain: A meta-analysis* [Unpublished PhD]. University of Florida, FL.

Fisher, D., & Frey, N. (2015). *Checking for understanding: Formative assessment techniques for your classroom* (2nd ed.). ASCD.

Fisher, D., Frey, N., Anderson, H., & Thayre, M. (2015). *Text-dependent questions: Pathways to close and critical reading.* Corwin.

Fisher, D., Frey, N., & Hattie, J. (2016). *Visible learning for literacy, grades K–12: Implementing the practices that work best to accelerate student learning.* Corwin.

Frayer, D., Frederick, W. C., & Klausmeier, H. J. (1969). *A schema for testing the level of cognitive mastery.* Center for Education Research.

Gladwell, M. (2008). *Outliers: The story of success.* Little, Brown, and Co.

Gravemeijer, K., & Doorman, M. (1999). Context problems in realistic mathematics education: A calculus course as an example. *Educational Studies in Mathematics, 39*(1/3), 111–129.

Hattie, J. (2009). *Visible learning: A synthesis of over 800 meta-analyses relating to achievement.* Routledge.

Hattie, J. (2023). *Visible learning: The sequel: A synthesis of over 2,100 meta-analyses relating to achievement.* Routledge.

Houser, M. L., & Frymier, A. B. (2009). The role of student characteristics and teacher behaviors in students' learner empowerment. *Communication Education, 58*(1), 35–53. https://doi.org.libproxy.sdsu.edu/10.1080/03634520802237383

Jackson, K., Garrison, A., Wilson, J., Gibbons, L., & Shahan, E. (2013). Exploring relationships between setting up complex tasks and opportunities to learn in concluding whole-class discussions in middle-grades mathematics instruction. *Journal for Research in Mathematics Education, 44*(4), 646–682. https://doi.org/10.5951/jresematheduc.44.4.0646

Jacobs, V. R., Lamb, L. L. C., & Philipp, R. A. (2010). Professional noticing of children's mathematical thinking. *Journal for Research in Mathematics Education, 41*(2), 169–202. www.jstor.org/stable/20720130

Kagan, S., & Kagan, M. (2021). *Kagan cooperative learning*. Kagan.

Kilpatrick, J., Swafford, J., & Findell, B. (Eds.). (2001). *Adding it up: Helping children learn mathematics*. National Academies Press.

Moll, L. C. (1990). Vygotsky's zone of proximal development: Rethinking its instructional implications. *Journal for the Study of Education and Development, 13*(51–52), 157–168.

National Council of Teachers of Mathematics. (2014). *Principles to actions: Ensuring mathematical success for all*. NCTM.

National Governors Association Center for Best Practice, Council of Chief State School Officers. (2010). *Common core state standards for mathematics*. Author.

Nicol, D. J., & Macfarlane-Dick, D. (2006). Formative assessment and self-regulated learning: A model and seven principles of good feedback practice. *Studies in Higher Education, 31*(2), 199–218. https://doi.org/10.1080/03075070 600572090

Rasmussen, C., Yackel, E., & King, K. (2003). Social and sociomathematical norms in the mathematics classroom. In H. Schoen & R. Charles (Eds.), *Teaching mathematics through problem solving: Grades 6-12* (pp. 143–154). National Council of Teachers of Mathematics.

Reinholz, D. L. (2015). Peer-assisted reflection: A design-based intervention for improving success in calculus. *International Journal for Research in Undergraduate Mathematics Education, 1*, 234–267.

Russell, B. (1945). *A history of western philosophy*. Simon and Schuster.

San Francisco Unified School District Mathematics Department. (2024). *Three-read protocol*. San Francisco Unified School District. https://www .sfusd.edu/departments/mathematics-depart ment-page/math-teaching-toolkit/mathteaching-strategies/signature-strategies/three-read-protocol

Serki, N., & Bolkan, S. (2024). The effect of clarity on learning: Impacting motivation through cognitive load. *Communication Education, 73*, 29–45. https://doi.org/10.1080/03634523.202 3.2250883

Titsworth, S., Mazer, J. P., Goodboy, A. K., Bolkan, S., & Myers, S. A. (2015). Two meta-analyses exploring the relationship between teacher clarity and student learning. *Communication Education, 64*(4), 385–418. https://doi.org.libproxy.sdsu .edu/10.1080/03634523.2015.1041998

Turner, E., Dominguez, H., Maldonado, L., & Empson, S. (2013). English learners' participation in mathematical discussion: Shifting positionings and dynamic identities. *Journal for Research in Mathematics Education, 44*(1), 199–234. https:// doi.org/10.5951/jresematheduc.44.1.0199

Vygotsky, L. S. (1978). *Mind in society: The development of higher psychological processes* (M. Cole, V. John-Steiner, S. Scribner, & E. Souberman Eds.; A. R. Luria, M. Lopez-Morillas, M. Cole, & J. V. Wertsch, Trans.). Harvard University Press.

Wong, H. M. (2014). I can assess myself: Singaporean primary students' and teachers' perceptions of students' self-assessment ability. *Education 3–13, 44*, 442–457. https://doi.org/10.1080/03004279 .2014.982672

Yackel, E., & Cobb, P. (1996). Sociomathematical norms, argumentation, and autonomy in mathematics. *Journal for Research in Mathematics Education, 27*(4), 458–477. https://doi.org/ 10.5951/jresematheduc.27.4.0458

Yackel, E., & Rasmussen, C. (2002). Beliefs and norms in the mathematics classroom. In G. Leder, E. Pehkonen, & G. Toerner (Eds.), *Beliefs: A hidden variable in mathematics education?* (pp. 313–330). Kluwer.

Yoon, B. (2007). Offering or limiting opportunities: Teachers' roles and approaches to English-language learners' participation in literacy activities. *The Reading Teacher, 61*, 216–225.

Yoon, B. (2008). Uninvited guests: The influence of teacher's roles and pedagogies on the positioning of English language learners in the regular classroom. *American Educational Research Journal, 45*, 495–522.

Zimmerman, B. J. (2002). Becoming a self-regulated learner: An overview. *Theory Into Practice, 41*(2), 64–70. https://doi.org/10.1207/s15430421tip 4102_2

Index

CORWIN Mathematics

Supporting TEACHERS | *Empowering* STUDENTS

PETER LILJEDAHL

14 optimal practices for thinking that create an ideal setting for deep mathematics learning to occur.

Grades K–12

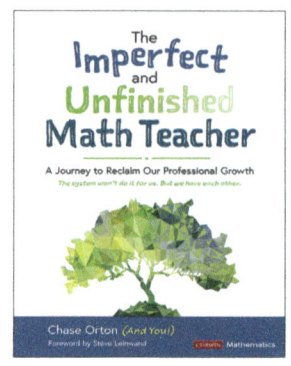

CHASE ORTON

A guide that leads math teachers through a journey to cultivate a more equitable, inclusive, and cohesive culture of professionalism for themselves.

Grades K–12

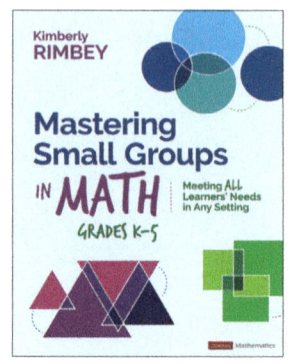

KIMBERLY RIMBEY

Much needed guidance on how to meet the diverse needs of students using small group math instruction.

Grades K–5

BETH MCCORD KOBETT, FRANCIS (SKIP) FENNELL, KAREN S. KARP, DELISE ANDREWS, LATRENDA KNIGHTEN, JEFF SHIH, DESIREE HARRISON, BARBARA ANN SWARTZ, SORSHA-MARIA T. MULROE

Detailed plans for helping elementary students experience deep mathematical learning.

Grades K–1, 2–3, 4–5

LOU EDWARD MATTHEWS, SHELLY M. JONES, YOLANDA A. PARKER

A resource for designing inspiring learning experiences driven by the kind of high-quality and culturally relevant mathematics tasks that connect students to their world.

Elementary, Middle and High School

JOHN J. SANGIOVANNI, SUSIE KATT, KEVIN J. DYKEMA

A guide for empowering students to embrace productive struggle to build essential skills for learning and living—both inside and outside the classroom.

Grades K–12

To order, visit corwin.com/math

**JENNIFER M. BAY-WILLIAMS,
JOHN J. SANGIOVANNI,
ROSALBA SERRANO,
SHERRI MARTINIE,
JENNIFER SUH, C. DAVID WALTERS**

Because fluency is so much more
than basic facts and algorithms.
Grades K–8

**JOHN J. SANGIOVANNI, SUSIE KATT,
LATRENDA D. KNIGHTEN,
GEORGINA RIVERA,
FREDERICK L. DILLON,
AYANNA D. PERRY,
ANDREA CHENG, JENNIFER OUTZS**

Actionable answers to your most
pressing questions about teaching
elementary and secondary math.

Elementary, Secondary

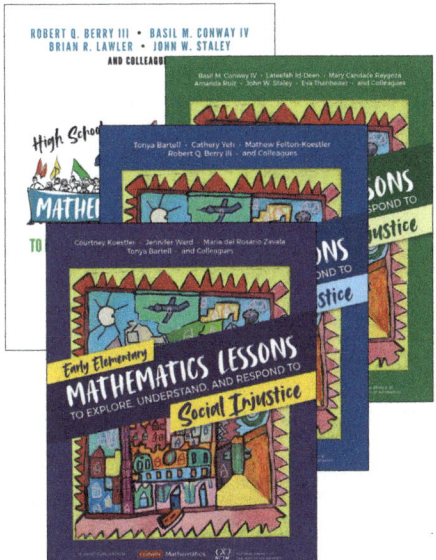

**ROBERT Q. BERRY III, BASIL M. CONWAY IV,
BRIAN R. LAWLER, JOHN W. STALEY,
COURTNEY KOESTLER, JENNIFER WARD,
MARIA DEL ROSARIO ZAVALA,
TONYA GAU BARTELL, CATHERY YEH,
MATHEW FELTON-KOESTLER,
LATEEFAH ID-DEEN,
MARY CANDACE RAYGOZA,
AMANDA RUIZ, EVA THANHEISER**

Learn to plan instruction that engages
students in mathematics explorations
through age-appropriate and culturally
relevant social justice topics.

**Early Elementary, Upper Elementary,
Middle School, High School**

**SARA DELANO MOORE,
KIMBERLY RIMBEY**

A journey toward making
manipulatives meaningful.
Grades K–3, 4–8

CORWIN

A Sage Company

Helping educators make the greatest impact

CORWIN HAS ONE MISSION: to enhance education through intentional professional learning.

We build long-term relationships with our authors, educators, clients, and associations who partner with us to develop and continuously improve the best evidence-based practices that establish and support lifelong learning.